Fiscal Policy to Mitigate Climate Change

A Guide for Policymakers

EDITORS
Ian W.H. Parry, Ruud de Mooij, and Michael Keen

INTERNATIONAL MONETARY FUND

Cataloging-in-Publication Data
Joint Bank-Fund Library

Fiscal policy to mitigate climate change : a guide for policymakers / Ruud A. de Mooij, Ian W.H. Parry, and Michael Keen. – Washington, D.C. : International Monetary Fund, 2012.
p. ; cm.

Includes bibliographical references.
ISBN 978-1-61635-393-3

1. Climatic changes – Government policy. 2. Greenhouse gas mitigation – Government policy. 3. Carbon dioxide mitigation – Economic aspects. 4. Carbon taxes. Carbon dioxide – Taxation. 6. Emissions trading. I. Mooij, Ruud A. de. II. Parry, Ian W. H. (Ian William Holmes), 1965- III. Keen, Michael. IV. International Monetary Fund.

QC981.8.C5.F57 2012

Please send orders to:
International Monetary Fund, Publication Services
P.O. Box 92780, Washington, DC 20090, U.S.A.
Tel.: (202) 623-7430
Fax: (202) 623-7201
E-mail: publications@imf.org
Internet: www.imfbookstore.org

Contents

Foreword

Global warming poses critical policy challenges, now and for the coming years, with potentially profound implications for macroeconomic performance and economic well-being.

These challenges are, to an important degree, ones for the design of national tax and spending systems. The world needs strategies for adapting to the medium- and long-term consequences of climate change, and these have fiscal implications. Most pressing, however, is the need for appropriate policies to "mitigate"—that is, to limit—greenhouse gas (GHG) emissions. This need is very widely acknowledged, although the appropriate scale of (near-term and longer-term) mitigation policies, as well as the responsibilities of developing countries, remain contentious. Without significant emissions reductions, most studies project global temperature rises of 2.5° C to 6.5° C above preindustrial levels by the end of this century. The associated uncertainties and risks are substantial.

Tax and similar pricing instruments have a crucial role to play in this area. We need to understand both their environmental effectiveness and their impacts on competitiveness, different household groups, and overall fiscal positions.

This volume seeks to provide policymakers with practical guidelines for the design and implementation of climate mitigation policies. The premise at its heart is that fiscal instruments—carbon taxes or their cap-and-trade equivalents (with auctioned allowances)—can and should form the centerpiece of policies to reduce energy-related carbon dioxide emissions (which account for about 70 percent of projected GHG emissions). These pricing policies can also become a large new source of government revenue, which could make a significant contribution to meeting fiscal consolidation challenges and, more generally, to building more efficient and fairer tax systems. So addressing climate change can be both a challenge and an opportunity.

As the contributions in this volume make clear, this premise reflects what is (for once) a strong consensus among economists. Important areas of disagreement remain, of course, and the various chapters explore many of these. But there is wide agreement on the central role that fiscal instruments must play if we are to address climate change effectively and efficiently. I hope that the guidelines set out in this volume, prepared by some of the leading experts in the field, will contribute not only to informed debate but also to much-needed action.

Christine Lagarde
Managing Director
International Monetary Fund

Summary for Policymakers

Ruud de Mooij, Ian Parry, and Michael Keen
Fiscal Affairs Department, International Monetary Fund

Scientific evidence suggests that climate change is an extremely serious threat (see Box I.1) and that a major international effort to slow atmospheric accumulations of greenhouse gases (GHGs) over the twenty-first century is a key component of the appropriate policy response. If left unchecked, climate change could have increasingly serious macroeconomic consequences—especially in countries with limited ability to adapt to hotter temperatures, higher sea levels, diminished water supplies, and so on.

Many countries have made emissions control pledges, and parties at the December 2011 climate change meetings in Durban, South Africa, pledged to develop a global emissions control agreement to be implemented in 2020. However, until there are credible mechanisms for enforcing such commitments, it is not entirely clear that the Durban platform will deliver on its promise, in which case climate policies will continue to emerge in a piecemeal "bottom up" fashion for the foreseeable future. Either way, the implementation of mitigation policy is only just beginning: Over 90 percent of global GHG emissions are presently not covered by formal mitigation programs.

In responding to this challenge, it is critical to use the most *effective* emissions control instruments, namely those that exploit all potential possibilities for reducing emissions, rather than using narrowly focused policies that miss out on a lot of these opportunities. It is also important to use policies that contain mitigation *costs* (for a given emissions reduction), not only for its own sake, but also to improve the prospects for sustaining policies over time.

The instrument that best fits these two criteria is revenue-raising carbon pricing—carbon taxes or cap-and-trade systems with allowance auctions—so long as it is well designed in terms of comprehensively

Box I.1. The Climate Change Challenge

Annual global CO_2 emissions from fossil fuels have grown from about 2 billion tonnes in 1900 to about 30 billion tonnes today, and, in the absence of mitigation policies, they are projected to roughly triple 2000 levels by the end of the century. The huge bulk of the projected future emissions growth is in developing countries: CO_2 emissions from these countries now exceed those from industrial countries; by 2030, China and India combined are expected to account for about one-third of global emissions. Land-use changes (primarily deforestation) will contribute about an additional 5.5 billion tons of CO_2 releases, though these sources are projected to grow at a much slower pace than fossil fuel emissions.

Atmospheric CO_2 concentrations have increased from preindustrial levels of about 280 parts per million (ppm) to current levels of approximately 390 ppm, and they are projected to rise to about 700 to 900 ppm by 2100. About one-half of CO_2 releases accumulate in the atmosphere (the rest are absorbed by sinks, especially the oceans and forests). Accounting for non-CO_2 GHGs, such as methane and nitrous oxides, the CO_2-*equivalent* atmospheric concentration is about 440 ppm. Total GHG concentrations in CO_2-*equivalents* are projected to reach 550 ppm (i.e., about double preindustrial levels) by around mid-century.

The globally averaged surface temperature is estimated to have risen by about 0.75° C since 1900, with most of this warming due to rising GHG concentrations. If CO_2-*equivalent* concentrations were stabilized at 450, 550, and 650 ppm, mean projected warming over preindustrial levels is 2.1° C, 2.9° C, and 3.6° C, respectively, once the climate system stabilizes (which takes several decades). Actual warming may exceed (or fall short of) these projections due to poorly understood feedback in the climate system.

The physical consequences of warming include changed precipitation patterns, sea level rise (amplified by storm surges), more intense and perhaps frequent extreme weather events, and possibly more catastrophic outcomes like runaway warming, melting of ice sheets, or destruction of the marine food chain (due to warmer, more acidic oceans). Estimates of the damages from these effects are highly uncertain due to difficulties in valuing low-probability, catastrophic events; uncertainty over regional climate effects (including the risk of shifting monsoons and deserts); and uncertainty over regional development, technological change (including adaptive technologies like climate and flood-resistant crops), and other policies (e.g., attempts to eradicate malaria or integrate global food markets). Worldwide impacts also mask huge disparities in regional burdens—hotter, low-lying, and low-income countries are most at risk and are most lacking in adaptive capability, while some wealthy, more temperate countries could benefit (e.g., from longer growing seasons).

Sources: Chapter 3, IPCC (2007), and Aldy and others (2010).

covering emissions. Revenues from these fiscal instruments can contribute significantly to fiscal consolidation needs—if countries do not implement such policies, they will need to rely more heavily on other deficit reduction measures.

However, policymakers may need to consider many questions in crafting carbon pricing legislation. These include the following:

- How strong is the case for carbon pricing instruments over regulatory approaches (e.g., standards for energy efficiency or mandates for renewables), how do carbon taxes and cap-and-trade systems compare, and what might be some promising alternatives if "ideal" pricing instruments are not viable initially?

- How is a carbon pricing system best designed in terms of covering emissions sources, using revenues, overcoming implementation obstacles (e.g., by dealing with competitiveness and distributional concerns), and possibly combining them with other instruments (e.g., technology policies). And how might pricing policies be coordinated across different countries?

- How should policymakers think about the appropriate level of emissions pricing?

- How important is inclusion of the forest sector in carbon pricing schemes, and how feasible is this in practice?

- What should be the priorities for developing economies in terms of fiscal reforms to reduce emissions?

- From the perspective of raising funds (from developed economies) to fund climate projects (in developing economies), what are the most promising fiscal instruments and how should they be designed?

- What lessons can be drawn from experience with emissions pricing programs, like the European Emissions Trading System (ETS), or the various carbon tax programs to date?

Although the IMF is not an environmental organization, environmental issues matter for our mission when they have major implications for macroeconomic performance and fiscal policy. Climate change clearly passes both these tests, and in fact recent IMF work has addressed a variety of fiscal issues posed by climate and broader environmental challenges.[1] Continuing this work, in September 2011 the IMF's Fiscal Affairs Department held an expert workshop at which eight policy notes covering the above, and some other, issues in designing carbon pricing policies were presented for discussion and comment. This volume collects the final versions of these policy briefs.

To be sure, there are numerous other discussions on the design of climate mitigation policies. However, this volume differs because of its especially in-depth coverage of issues for *fiscal* policies and provision of specific, readily implementable, policy recommendations. Other components of the appropriate response to climate change, including adaptation policy, improving scientific knowledge, and developing "last-resort" technologies for use in extreme climate scenarios, are beyond our scope.[2]

The rest of this summary draws out some of the main take-home lessons for policymakers from the different chapters in this volume. Most of these lessons are actually fairly straightforward—climate policy design is not as complicated as it might first appear.

Lessons from Chapter 1: Comprehensive Carbon Pricing Policies Can Effectively Reduce Emissions and at Least Cost

Comprehensive carbon pricing measures exploit the entire range of emissions reduction opportunities across the economy. As the emissions price is reflected in the prices of fossil fuels, electricity, and so on, this promotes fuel switching in the power sector and reductions in the demand for electricity, transportation fuels, and direct fuel usage in homes and industry. Carbon pricing also strikes the cost-effective balance between different emission reduction opportunities because all behavioral responses are encouraged up to where the cost of the last tonne reduced equals the emissions price. Moreover, the carbon price provides a strong signal for innovations to improve energy efficiency and reduce the costs of zero- or low-carbon technologies. By definition, regulatory policies on their own, like mandates for renewable fuel generation and energy efficiency standards,

[1] This work covers, for instance, the macroeconomic, fiscal, and financial implications of climate mitigation and adaptation policies; the appropriate design of fuel and other environmental taxes; the measurement of energy subsidies and protection of the poor when they are scaled down; border tax adjustments; and the taxation of resource industries. For more information see www.imf.org/external/np/exr/facts/enviro.htm.

[2] "Last-resort" technologies include "air capture" filters to absorb CO_2 from the atmosphere and store it underground (at present, these technologies are unproven and very costly). They also include "geo-engineering" technologies, like solar radiation management (shooting particulates into the atmosphere to deflect incoming sunlight), which are inexpensive to deploy and could entail dangerous downside risks (e.g., the possibility of overcooling the planet).

are far less effective as they focus on a much narrower range of emission reduction opportunities. Regulatory policies can also impose excessive costs unless they are accompanied by provisions allowing firms with high emissions control costs to purchase emission reduction credits from firms with low emission control costs. Given the scale of the challenge—reducing emissions to a minor fraction of "business-as-usual" emissions over coming decades—choosing the most effective and cost-effective mitigation instruments is critically important.

The choice between carbon taxes and emissions trading systems is generally less important than implementing one of them and getting the design details right. Key design specifics include comprehensively covering emissions and avoiding the squandering of revenue potential (e.g., by granting free allowance allocations in cap-and-trade systems or earmarking revenues for socially unproductive purposes). For cap-and-trade systems, provisions are also needed to limit price volatility, and these systems are not appropriate for countries lacking institutions to support credit trading.

If carbon pricing policies are not initially viable, carefully designed regulatory packages or, better still, "feebates" can be reasonable alternatives. Combining a carbon dioxide (CO_2) per kilowatt hour standard for the power sector with energy-efficiency standards for vehicles, appliances, buildings, and so on can promote many of the emission reduction opportunities that would be exploited by carbon pricing policies. And regulatory policies avoid large (politically challenging) increases in energy prices as they do not involve the pass-through of large carbon tax revenues (or allowance rents) in higher prices. But again, extensive credit-trading provisions across firms and sectors are important for containing the costs of these regulatory packages. More promising is to use feebate or tax/subsidy analogs of these regulations (e.g., taxes for generators with high emissions intensity and subsidies for generators with low emissions intensity), as these policies circumvent the need for credit trading. Regulatory or feebate policy packages should still transition to carbon pricing whenever feasible, however, to raise government revenue, more comprehensively reduce emissions, and facilitate international coordination.

Lessons from Chapter 2: Design Details for Carbon Pricing Are Important

Targeting the right base for carbon pricing is critical for environmental effectiveness. Ideally, carbon prices are applied in proportion to the carbon content of fuels as they enter the economy (e.g., from petroleum refineries, coal mines, fuel importers), with refunds for carbon capture technologies installed at industrial facilities. Pricing the carbon content of fuels at different rates, or varying the price across fuel users, undermines cost-effectiveness by placing an excessive burden of reductions on the heavily taxed fuel or end user and too little burden on other fuels or other users. Electricity taxes are a poor substitute for carbon pricing on environmental grounds, as the former miss out on the huge bulk of emissions mitigation opportunities. Pricing emissions at the point of fuel combustion (e.g., power generators, industrial boilers) involves monitoring many more entities and some loss in coverage (e.g., small-scale emitters are usually exempt). Some non-CO_2 GHGs might be covered directly under the pricing regime or indirectly through emissions offset credits, as capability for monitoring and verification is developed over time.

The costs of comprehensive carbon pricing is initially modest if revenues are used productively. Productive revenue uses include reducing taxes on work effort and capital accumulation, retiring public debt, and funding socially desirable (environmental or other) public spending. With productive revenue use, the overall costs of (appropriately scaled) carbon taxes to the economy are modest in the medium term, typically around 0.03 percent of GDP for developed economies. If revenues are squandered, however, policy costs can be several times higher.

Although carbon pricing is the most important measure for promoting clean technology development and deployment, supplementary technology policies may be warranted, though they need to be carefully designed. For example in cases where, despite carbon pricing, clean technology deployment could be too slow because of further "market failures," additional transitory incentives may be appropriate. Pricing incentives (e.g., technology adoption subsidies) are generally better able to handle uncertainty over future technology costs than technology mandates that force a technology, regardless of future conditions.

Some options for overcoming opposition to carbon pricing do exist. Higher energy prices hurt consumers and reduce the competitiveness of trade-exposed, energy-intensive firms (e.g., aluminum and steel producers). However, these effects should not be overstated and might be addressed in part through scaling back preexisting energy taxes (particularly on vehicles and electricity consumption) that become redundant with carbon pricing. Another possibility is to compensate through the broader fiscal system (e.g., in Australia, revenues from carbon pricing will fund an increase in personal income tax thresholds to especially help low-income households). Competitiveness concerns might be addressed through transitory production subsidies for vulnerable firms (this is better than granting these sectors preferential fuel prices). Border tax adjustments are another possibility, though they may run afoul of free trade obligations.

At an international level, a price floor among large emitting countries is a potentially promising way forward. Reaching an international agreement over a common CO_2 price and how it might respond to future evidence on global warming may be less difficult than agreeing on annual emissions targets for each participating country. Prospects for agreement might be enhanced further if the policy took the form of a floor price, which provides some protection for countries willing to set relatively higher carbon prices.[3] Although countries would forgo controls over annual emissions, pricing agreements might be combined with maximum allowable emissions cumulated over, say, a 10-year period (requiring increases in their carbon price if they are not on track to stay within the "carbon budget"). An agreement would need provisions (e.g., monitoring by an international body) to deal with the possibility that individual countries may adjust their broader energy tax/subsidy provisions to undermine some of the effectiveness of the formal carbon price.

Lessons from Chapters 3 and 4: Studies Suggest that a Reasonable Starting Level for Emissions Prices in Large Emitting Countries Would Be about US$20 Per Tonne of CO_2 or More by 2020

There are two basic ways to think about the appropriate price on CO_2 (and other GHGs). One is to define an ultimate goal for global climate stabilization—usually a target for mean projected warming above the pre-industrial level (that might, for example, be the result of a political process, of an ethical principle, or of a precautionary approach)—and impose emissions pricing paths that are consistent with meeting this target, ideally in a way that minimizes mitigation costs. The other is to impose emissions prices that reflect potential environmental damages per tonne of emissions. Despite considerable uncertainties and controversies, broad policy guidance can still be provided under either paradigm.

Limiting long-term, mean projected warming to 2° C above preindustrial levels—the official goal of the UN Framework Convention on Climate Change—is highly ambitious and may be infeasible. As indicated in Figure I.1, atmospheric concentrations of GHGs would need to be stabilized at about 450 parts per million (ppm) of CO_2 equivalent (or close to current levels) in order to keep projected warming to 2° C. After an inevitable period of "overshooting" this concentration level, global GHG emissions would need to be negative on net for a sustained future period to bring CO_2 equivalent

[3] Alternatively, a common floor price might be agreed among countries implementing cap-and-trade systems, without necessarily any agreement over country-level emissions caps.

Figure I.1. Projected Long-Term Warming above Preindustrial Temperatures from Stabilization at Different Greenhouse Gas Concentrations

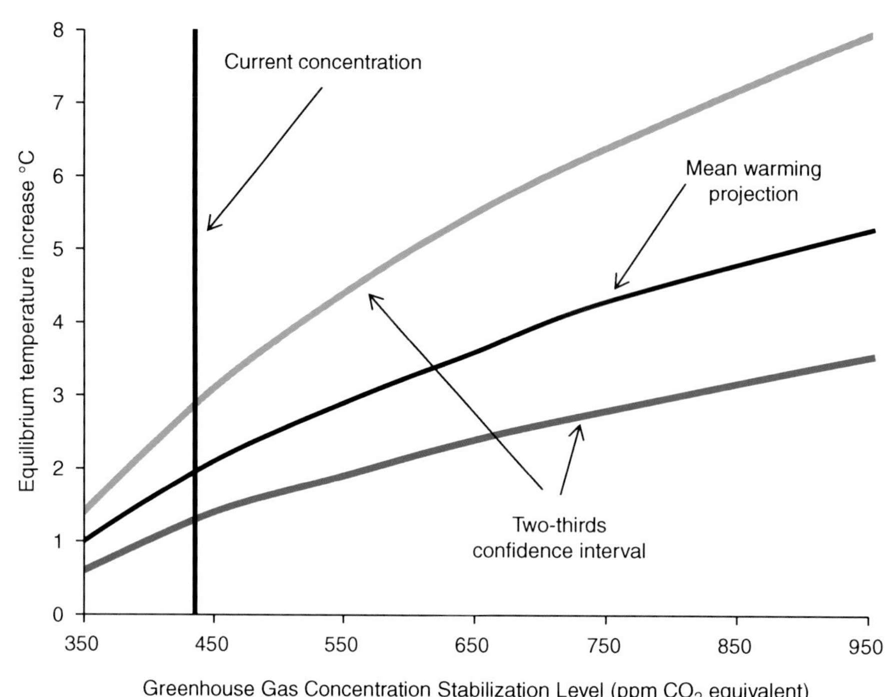

Source: IPCC (2007), Table 10.8.

concentrations back down to 450 ppm. Whether negative emission technologies (e.g., use of biomass in power generation coupled with carbon capture and storage) can be developed, let alone deployed on a global scale, to more than offset other GHG emissions, is highly speculative.

Less stringent targets, for example, keeping mean projected warming to 2.9° C or 3.6° C, are more plausible, but also more risky. These warming targets require stabilizing atmospheric GHG concentrations at approximately 550 or 650 ppm of CO_2 equivalent, respectively, and global emissions prices in the ballpark of US$40 and US$20 per tonne by 2020, respectively. Relative to business-as-usual outcomes, these stabilization targets substantially reduce the risk of more extreme climate outcomes—but this risk is not eliminated (underscoring the need for investment in last-resort technologies). Postponing mitigation actions, especially in emerging economies, can greatly raise the global costs of climate stabilization and render more stringent targets infeasible. For example, the 550-ppm target becomes technically out of reach if action by all countries is delayed beyond 2030.

Encouraging the major emitting developing economies to reduce GHGs could be facilitated by compensation payments. This might take the form of direct side payments (under a tax regime) or generous emissions allocations (under a trading system), though both are challenging to negotiate. In the meantime, the Green Climate Fund (GCF) could catalyze financial flows to developing economies, underscoring the need for innovative sources to finance the GCF.

As policies emerge piecemeal (rather than as part of an internationally agreed stabilization goal), it might be more natural to base the emissions price on the social cost of carbon (SCC). The SCC is the discounted monetary value of the future climate change damages due to an additional tonne of CO_2 emissions. A recent U.S. government study (in their central case) recommended a value of $21.4 per tonne of emissions released in 2010 (in 2007 U.S. dollars), rising at about 2 to 3 percent per year in real terms (i.e., this price is roughly consistent with near-term prices for stabilizing projected warming at 3.6° C). These estimates are based on an extensive assessment of models combining simplified representations of the climate system with dynamic models of the global economy. Damages reflect, for example, future impacts on world agriculture, costs of protecting against rising sea levels, health effects (e.g., from heat waves), ecological impacts (e.g., species loss), and risks of more extreme damage outcomes.

The SCC is sensitive, in particular, to alternative perspectives on discounting and extreme climate risks. Global warming is an intergenerational problem because emissions have long atmospheric residence times (about 100 years in the case of CO_2), and the full warming from higher atmospheric concentrations is not felt for several decades (due to gradual heat diffusion processes in the oceans). Reports by the U.K. and German governments, for example, have cautioned against discounting impacts on future (unborn) generations on ethical grounds, in which case the SCC is much higher. The SCC can also be much higher if more weight is attached to the risk of extreme climate outcomes or if impacts on low-income countries are given a disproportionately high weight. Nonetheless, individual countries may be reluctant to price emissions much above US$25 per tonne, in the absence of similar pricing by other countries.

SCC values can be applied to other major emitting countries based on purchasing power parity exchange rates. Ideally, emissions from different countries should be priced at the same rate as they cause the same damage. Arguments can be made for exempting (low-emitting) developing economies (see below), for example if compensation payments (from wealthier countries) are not feasible.

Given the scope for future learning about the seriousness of climate change, establishing emissions pricing in the high-emitting countries over the next several years is more important at this stage than negotiating over long-range targets. Once pricing policies have been established, they can be adjusted as needed in the future as greater consensus emerges on the urgency of climate stabilization. A reasonable minimum price to aim for seems to be around US$20 per tonne, under either least-cost climate stabilization or damage valuation approaches. Establishing a credible time path for progressively rising carbon prices is also important to create stable incentives for long-term, clean energy investments.

Lessons from Chapter 5: National Payments for Forest Carbon Sequestration Are Promising if Carbon Can Be Measured

Potentially, forest carbon sequestration could account for about a quarter of global CO_2 mitigation over this century. This carbon storage could be achieved through a combination of reduced deforestation, afforestation, and changes in forest management, primarily in tropical regions.

Although a host of small programs can promote forest carbon sequestration, national programs are easier to administer. Scaling up small projects is difficult, given limited technical capacity (e.g., that of NGOs) and the risk of leakage (e.g., reduced deforestation in one area offset by increased deforestation elsewhere). Moreover, judging whether sequestration projects are "additional" (i.e., whether they would have gone ahead even without the program incentive) can be difficult (implying the possibility of "wasted" program funds). National programs are more promising ideally with coordination (e.g., harmonized emissions prices) among the major tropical forest countries (national programs also allow flexibility for governments to deal with multiple claimants over forest land).

Baselines need to be carefully chosen, however. One (national-level) possibility is tax-subsidy schemes (with periodic updating of baselines). For example, payments could be offered for increases in sequestered carbon on particular parcels of forest land over and above the sequestered amount in some baseline year, and charges could be applied to other parcels where sequestered carbon falls below the baseline level. The scheme would be approximately revenue neutral (even though some sequestration activities may not be additional). Measuring sequestered carbon can be difficult, however (e.g., it varies with species, tree age, and selective harvesting), and is not permanent. In cases where reliable estimates of carbon storage are lacking, payments may need to be based on some proxy for CO_2, like forest coverage, with adjustments for tree species and local climate. Ideally, payments are made on an annual basis to deter, for example, early harvesting and inadequate safeguards against fire hazards.

Lessons from Chapter 6: Small-Emitting Developing Economies Should Focus First on Energy Pricing Reforms that Are in Their Own Interest

Most low-income countries contribute very little to current and projected future CO_2 emissions and the case for them to undertake costly mitigation policies is correspondingly weak. Low-emitting developing economies nonetheless have a critical role in finding an effective and efficient global response to the challenges from climate change: ways need to be found both to prevent carbon leakage as mitigation measures in high-emitting countries cause emissions to shift there, and to exploit the relatively cheap opportunities for emissions reduction there. At least initially, emissions mitigation in low-income countries might be better promoted through climate finance (e.g., international offset programs and direct investments from climate funds).

As regards policy reform, low-income countries should focus on "getting energy prices right" from a local perspective, which would also have climate benefits. The first priority is to scale back any fossil fuel subsidies (especially consumer subsidies for high-carbon fuels). Although these subsidies are often rationalized on distributional grounds, such concerns are better addressed through more targeted policies (e.g., safety nets, investments in primary education), rather than artificially holding down energy prices (which benefits everyone, and often the rich more than the poor). The second priority is then to impose appropriate taxes on energy. Revenue considerations should involve integrating consumption of energy products under broader value-added tax systems. In fact, the case for taxation of energy is especially robust in developing economies, where problems of weak administration and tax compliance hinder the effectiveness of broader fiscal instruments. Further excise taxes on fuels are warranted to cover their potential local environmental damages (e.g., charges for the human health risks due to local pollution) and other side effects associated with fuel use (e.g., traffic congestion).

Lessons from Chapter 7: International Aviation and Maritime Charges Are a Promising Source of Climate Finance, although There Are Other Options

Pricing carbon for international aviation and maritime fuels is an appealing source of climate finance, as national governments do not have an obvious claim on the tax base. In addition, these fuels are currently under-taxed from both an environmental perspective and from a broader fiscal perspective (e.g., airline passenger tickets are exempt from value-added taxes), and they would be relatively straightforward to administer (e.g., on fuel distributors). The charges would need to be coordinated internationally to limit tax avoidance and competitiveness concerns. Compensation may be needed to secure the early participation of developing economies and to entice broader entry into the pricing agreement over time, but workable schemes should be feasible.

The case for carbon pricing and removal of fossil fuel subsidies is also strong. As already emphasized, carbon pricing should ideally form the centerpiece of mitigation efforts, and it could play a key role in catalyzing the private part of climate finance for developing economies (via emission offset programs). Removal of fossil fuel subsidies also provides mitigation benefits. These pricing reforms would yield substantial new revenues—potentially about US$250 billion a year for appropriately scaled carbon pricing in developed economies and US$40 to $60 billion a year from subsidy reform—though governments may be reluctant to hand over much domestic revenue for international purposes in the current fiscal environment.

Revenues can also be raised through broader fiscal instruments, though in this case costs are not offset by mitigation benefits. For personal income taxes, corporate income taxes, and value-added (or sales) taxes, the general recommendation is that exemptions and other tax preferences should be scaled back first, as this is usually a less distortionary way to raise revenue than increasing overall tax rates. Taxes on the financial sector are another possibility, though efficiency considerations favor taxing financial activities rather than (as usually proposed) financial transactions.

Lessons from Chapter 8: Carbon Pricing Programs Have Evolved over Time with Experience

Market-based mitigation policies implemented to date have performed reasonably well on effectiveness and cost-effectiveness grounds compared with regulatory approaches. Often, however, these gains have fallen somewhat short of their full potential, partly because actual designs have deviated from economically efficient designs, due in part to preferential treatments and exemptions (e.g., in Scandinavian countries, carbon tax rates vary considerably across end users). Market-based policies have also induced innovation in and adoption of emissions-reducing technologies (though again these gains are not always as large as expected). Carbon leakage effects to date have been relatively modest.

Emissions pricing programs often take the form of "hybrid" schemes that combine upstream and downstream elements and emissions taxes with emissions trading. For example, in Australia, large downstream emitters are covered under a cap-and-trade system, while more diffuse sources (e.g., home heating fuels and transportation fuels), are covered by taxes on fuel distributors. Although ideally these systems should transition to a single, comprehensive pricing instrument, they may perform reasonably well for the time being as they can still cover most energy-related CO_2 emissions. But provisions to limit price volatility for emissions trading systems may be required to harmonize, approximately, emissions prices across sectors (thereby promoting cost effectiveness).

Although the Kyoto Protocol sought to simultaneously control six GHGs by translating them into a common index of CO_2 equivalents, no existing program covers all these gases. For administrative ease, most programs focus solely on energy-related CO_2 emissions, though many programs are now beginning to transition to a more comprehensive coverage of gases as monitoring and verification capacity improves.

Price volatility has been a significant concern in trading systems to date (though experience is limited to developed economies). Cap-and-trade systems often limit price volatility through provisions for permit banking (allowing entities to save permits for later use when expected allowance prices are higher) and advance auctions (allowing entities to buy allowances at current prices for use in several years). Permit borrowing (which allows entities to use permits before their designated date) is more restricted, due to a fear that firms might default on owed allowances (though this does not seem to have been a problem). If these provisions work reasonably well, there is less need to transition to a carbon tax on price volatility grounds.

Revenues from carbon taxes and auctioned allowances have been used for reducing other taxes, compensating industries, offsetting regressive impacts on households, and promoting renewable and energy efficiency programs. Use of revenues for industry compensation has diminished over time, however, with greater appreciation of the value of forgone revenues and tendency to overcompensate (power producers reaped windfall profits in the early phases of the EU trading scheme due to free allowance allocations, but future ETS allowances will be largely auctioned). As we might recommend, some programs (e.g., in Australia) have addressed adverse effects on low-income households with progressive adjustments to the broader tax system.

Emissions "offset" provisions are a common means for reducing the cost of cap-and-trade programs. But the challenge is to ensure that the credited emissions reductions outside of the formal program can be measured and would not have occurred anyway (without the offset credit). Due to concerns about credibility, most programs impose limits on offsets, but newer approaches attempt to distinguish between more credible offsets (which are allowed) and less credible ones (which are rejected). Under a carbon tax, offsets are not needed to contain the emissions price, but if they are not used, untaxed sectors are left completely without control. In either system, offsets can be introduced over time (e.g., to promote financial flows to developing economies) as verification techniques improve.

The Role of Finance Ministries

To date, environmental ministries have been most involved in climate change discussions. A final lesson is that finance ministries need to be more actively involved in carbon pricing policy, given the significant amount of revenues at stake and that these instruments are a natural extension of existing fuel excise tax systems.

References and Suggested Readings

For further discussion on the design of climate mitigation policies, see the following:

Aldy, Joseph, Alan J. Krupnick, Richard G. Newell, Ian W. H. Parry, and William A. Pizer, 2010, "Designing Climate Mitigation Policy," *Journal of Economic Literature,* Vol. 48, pp. 903–34.

Nordhaus, William D., 2008, *A Question of Balance: Weighing the Options on Global Warming Policies* (New Haven, Connecticut: Yale University Press).

Stern, Nicholas, 2007, *The Economics of Climate Change: The Stern Review* (Cambridge, UK: Cambridge University Press).

For more details on the science of global warming, see:

IPCC, 2007. *Climate Change 2007: The Physical Science Basis.* Contribution of Working Group I to the Fourth Assessment Report of the Intergovernmental Panel on Climate Change (New York: Cambridge University Press).

CHAPTER

1 What Is the Best Policy Instrument for Reducing CO_2 Emissions?

Alan Krupnick
Resources for the Future, United States

Ian Parry
Fiscal Affairs Department, International Monetary Fund*

Key Messages for Policymakers

- Carbon pricing policies (carbon taxes and emissions trading systems) are easily the best instruments on the grounds of effectiveness, cost-effectiveness, and promoting clean technology investments.
- However, design details are important. Policies should be comprehensive, raise revenue, and be used in socially productive ways. Emissions trading systems also require fluid credit trading markets (i.e., a large number of market participants and institutions to enforce property rights) and price stability provisions.
- Carbon pricing policies can be challenging to implement, however, partly because of burdens on households and (trade-sensitive) industries. These burdens can be more severe than for other instruments.
- In the absence of carbon pricing, packages of regulations can be a reasonable (although not as good) alternative in the interim. However, they must be carefully designed to exploit, insofar as possible, mitigation opportunities across all sectors, and they require extensive credit trading to contain costs.
- Combining a series of "feebates" (tax/subsidy policies) may be more promising, as this circumvents the need for credit trading.
- Other policies in isolation (e.g., renewable mandates) are usually a poor substitute for carbon pricing or comprehensive regulatory/feebate packages.

* We are grateful to Joe Aldy, Terry Dinan, Michael Keen, Chris Moore, Richard Morgenstern, Andrew Stocking, and Tom Tietenberg for very helpful comments and suggestions.

Despite the failure of the U.S. Congress to pass cap-and-trade legislation to control greenhouse gas (GHG) emissions, worldwide and even U.S. attention to developing efficient and effective policies to mitigate climate change is not waning. At the 2011 climate change meetings in Durban, South Africa (COP-17), the participating parties agreed that by 2015, they would try to negotiate an international GHG emissions control regime to begin in 2020, including both developed and developing economies. However these negotiations play out, countries will need to implement specific policies to reduce emissions, especially fossil fuel carbon dioxide (CO_2), which account for about 70 percent of global GHGs. The appropriate choice of instrument, or instruments, to reduce CO_2 emissions is, however, a complex policy decision.

For one thing, there are all sorts of instruments that could be used, ranging from market-based instruments like carbon taxes and cap-and-trade systems, to vehicle fuel economy standards, emissions standards, and incentives for renewable fuels (see Box 1.1 for an explanation of the main options).

Box 1.1. Main Alternative Instruments for Mitigating CO_2 Emissions

Carbon taxes. Ideally, these taxes are applied upstream in the fossil fuel supply chain in proportion to the carbon content of fuels. Alternatively, they could be levied on CO_2 emissions released from major industrial smokestacks.

Cap-and-trade systems. These policies put a cap on emissions by requiring that covered firms hold permits for each tonne of (potential or actual) emissions. The government restricts the quantity of allowances, and trading among covered sources establishes a market price for allowances. Again, these policies could be applied upstream to the carbon content of fuels or at the point of emissions releases.

Excise taxes on individual fuels (e.g., coal), electricity, or vehicles.

Energy efficiency standards. Applied to vehicles, these policies set minimum requirements on the average fuel economy (kilometers per liter) of vehicles sold by different firms or (almost equivalently) a maximum rate for average CO_2 per kilometer across vehicle sales. Ideally, credit trading would allow some producers (specializing in large vehicles) to fall short of the standard by purchasing credits from others that go beyond the standard. Standards can also be applied to improve the energy efficiency of new buildings, household appliances, and other electricity-using durable goods.

Emissions standards. For the power sector, this policy imposes a ceiling on the maximum allowable CO_2 per kilowatt hour (kWh), averaged across each generator's plants. Again, flexibility can be provided by allowing emissions-intensive generators to fall short of the standard by purchasing credits from other generators that go beyond the standard.

Box 1.1. (*continued*)

Incentives for renewable fuels. Policies to promote generation from renewables include renewable portfolio standards (minimum shares for renewables in a generator's fuel mix), subsidies for renewable generation, and feed-in tariffs (guaranteed prices for renewable generation).

Feebates. For vehicle sales, feebates apply fees to new vehicles in proportion to the difference between their CO_2 per kilometer and a "pivot point" level and corresponding rebates (or subsidies) to vehicles with CO_2 per kilometer below the pivot point. Similarly, in the power sector, feebates impose a per-kWh charge on generators in proportion to any difference between their average CO_2 per kWh and the pivot point and a rebate to generators with CO_2 per kWh below the pivot point. Feebates can be designed to raise some revenue, or be revenue neutral, depending on whether the pivot point is below or at the industry average emission rate.

Regulatory combinations. These involve a set of independent regulations designed to exploit many of the emission-reduction opportunities that would be exploited under comprehensive emissions pricing. For example, the combination might include an emissions standard for the power sector and various standards for the energy efficiency of vehicles and electricity-using durables. Alternatively, the feebate analogs to these regulations might be combined in a policy package.

Source: Authors.

Moreover, policymakers may be concerned about multiple criteria, including the following:

- *Effectiveness* in terms of reducing CO_2 emissions in the near term.
- *Economic costs*—a *cost-effective* policy is one that minimizes the burden on the economy from a given emissions reduction (accounting for use of any government revenues raised).
- *Ability to deal with uncertainty* over future fuel prices, the availability of emissions-saving technologies, and so forth.
- *Distributional impacts* across income groups and industries, which matter for fairness, competitiveness, and acceptability.
- *Promotion of clean technology development and deployment,* which matters for long-term effectiveness.

This chapter provides a framework for evaluating alternative policy instruments against the above criteria and understanding the potentially strong case for fiscal instruments (i.e., carbon taxes or their cap-and-trade equivalents with allowance auctions). The following five sections take each of the above criteria in turn, and a summary matrix at the end of the chapter ranks all the policies against the different criteria. The discussion mostly draws on insights from the economics literature on instrument choice (see Suggested Readings).

For clarity, policies are compared (approximately) for the same (explicit or implicit) price they place on CO_2 emissions or the same impact they have on energy prices. For example, when an electricity tax is compared with an economy-wide CO_2 tax, the policies are assumed to have about the same effect on electricity prices. This means that both policies can be cost-effective *for the emissions reductions they achieve,* but those reductions will be (much) larger under the CO_2 tax.

The discussion is not fully comprehensive. Many other policies are often rationalized on climate grounds (e.g., biofuel mandates or tax credits for hybrid vehicles), although their environmental effectiveness is typically on a smaller scale than the instruments considered here. And our criteria are not exhaustive: Policymakers may also care about the ease of negotiating international agreements and the development of international carbon markets (to facilitate financial and technology flows). The first is difficult to gauge, and in principle, all market-based and regulatory approaches could promote carbon markets through appropriate crediting provisions, though the market breadth will depend on the portion of domestic emissions covered by the mitigation instrument.[1]

Environmental Effectiveness

A policy's effectiveness depends on its ability to exploit possibilities for reducing (energy-related) CO_2 emissions across the economy. It is helpful to group the main possibilities into the following four categories:

- *Power sector fuel mix.* Reducing average CO_2 emissions per kilowatt hour (kWh) of power generation through switching from carbon-intensive fuels (coal) to less carbon-intensive fuels (natural gas, fuel oil) or zero-carbon fuels (nuclear, hydro, wind, solar, geothermal). Emissions intensity can also be reduced through technologies to improve plant efficiency

[1] Other possible criteria not considered here include administrative costs and the ease and accuracy of monitoring and enforcement (see Chapter 2 for some discussion on this topic). A further caveat is that the policies we discuss are broad-brush rather than finely detailed. Cap-and-trade systems implemented to date have involved considerable complexity (see Chapter 8), although the same may be true of other policies, such as carbon taxes, as they emerge from the legislative and regulatory process. Whether these details (e.g., on exempt sectors or earmarking of policy revenues) matter for the general conclusions drawn here would need careful study.

(i.e., reducing fuel requirements per kWh of generation). And carbon capture and storage (CCS) technologies may eventually prove viable in preventing CO_2 releases from fossil fuel plants.

- *Power sector output.* Reducing residential and industrial (including commercial) electricity demand through electricity-saving technologies (e.g., compact fluorescent lamps) as well as reduced use of electricity-using durables (e.g., economizing on the use of air conditioners).[2]
- *Direct non-electricity fuel use in homes and industry.* Reducing direct usage of fuels (e.g., natural gas) in homes, shops, factories, and offices.
- *Transportation fuels.* Reducing consumption of transportation fuels through reducing vehicle miles travelled and improving average vehicle fuel economy.

Market-Based Policies

Comprehensive (upstream) policies. A highly effective policy for reducing CO_2 emissions is a carbon tax applied upstream in the fossil fuel supply chain in proportion to the carbon content of each fuel (with refunds for any downstream capture of emissions by CCS). This tax system fully covers potential releases of CO_2 from later fuel combustion. To the extent the emissions tax is passed forward, it leads to higher prices for fossil fuels (especially coal, but also natural gas and petroleum products) as well as electricity. These higher energy prices encourage all of the above emission-reduction opportunities.

Cap-and-trade systems. These can be applied to the same base as the carbon tax and are therefore about equally effective over time. That is, as the value of allowances (i.e., the emissions price) is reflected in fuel and electricity prices, the policy will exploit the same emissions reduction opportunities as under the carbon tax.

Market-based policies with partial coverage (downstream). Another possibility is market-based policies focused at the point of emissions releases by large power and industrial plants. These policies are less effective at reducing emissions than upstream systems unless they are accompanied by measures to address transportation fuels, home heating fuels, and

[2] One caveat here is that electricity conservation tends to hit the most expensive (i.e., marginal) fuels first, which may be renewables or natural gas, rather than the highest carbon-emitting fuel, hence dampening the effect on emissions.

small-scale industrial sources. For example, by itself, the EU Emissions Trading Scheme covers about half of energy-related CO_2 emissions.[3]

Other energy taxes. Other energy taxes tend to be relatively ineffective at reducing CO_2 (see Chapter 2). Excise taxes on residential and industrial electricity use only exploit one of the four main emissions reduction opportunities.[4] Taxes on vehicle ownership are less effective still—even within the transport sector, they do not encourage people to drive a given vehicle less and may not (depending on how they are designed) create much demand for fuel-efficient vehicles. And while a coal tax is effective at reducing the most carbon-intensive fuel, it misses out on some opportunities exploited by a carbon tax, such as shifting from natural gas and fuel oil to nuclear and renewables and mitigation options outside of the power sector.

Direct Regulations

Regulatory policies by themselves can be expected to have (very) limited effects (particularly at the same implicit CO_2 price as the market-based instruments). These instruments need to be combined in far-reaching policy packages to achieve anything close to the effectiveness of comprehensive market-based policies. We distinguish among regulations focusing on increasing particular types of energy use (renewables), reducing carbon emissions, and reducing energy use.

Incentives for renewable generation. While there could be a rationale for transitory policies to promote renewables due to broader, technology-related market failures (see below), usually this is—or at least should be—as a *complement* to, not a *substitute* for, broader pricing instruments. These policies in isolation are not very effective relative to comprehensive pricing policies. They do nothing to reduce emissions outside of the power sector. At best, they only have weak incentives for electricity conservation as they do not involve the pass-through of carbon tax revenue or allowance value in higher generation prices.[5] And even within the power sector, they do

[3] Extending the EU emissions price to all emissions sources would not double emissions reductions, however. This is because most of the low-cost options for reducing CO_2 (for the European Union) are in the power sector or, put another way, emissions in the noncovered sector are less responsive to pricing than emissions that are already covered.

[4] These taxes are mandatory in the European Union under Energy Directive 2003/96/EC, although there are current discussions to revise this directive to target carbon emissions more directly.

[5] Under a renewable mandate, generators face higher average production costs per kWh because they shift away from their least-cost generation mix toward a cleaner, but more costly, generation mix. This also happens under market-based approaches applied at the point of emissions releases. In addition, however, average costs to generators, and hence generation prices, rise because generators must either pay a tax on their remaining CO_2 emissions per kWh or buy allowances to cover those emissions. In an upstream market-based system, carbon tax revenues or allowance value are already captured in the higher fuel prices paid by generators, which in turn are passed forward into electricity prices.

not exploit emissions reductions from replacing coal with natural gas and fuel oil or for switching from these fuels to nuclear.

Broader policies to decarbonize power generation. An industry-wide standard for CO_2 per kWh is a more effective approach than a renewables incentive policy because it encourages *all* possibilities for altering the generation mix to lower CO_2 emissions (not just substitution toward renewables) as well as improvements in plant efficiency. (As noted later, however, these types of regulatory policies need to be accompanied by extensive credit trading provisions to keep down their costs.) An emissions standard is closely related to the Clean Energy Standard, variants of which are currently under consideration in the United States. This policy sets minimum requirements on the share of zero-carbon fuels in power generation, but allows partial credits for fuels with intermediate carbon intensity.[6]

- There is also a pricing variant of the emissions standard, known as a feebate (see Box 1.1). This policy exploits the same incentives for reducing CO_2 per kWh as an emissions standard, but with some possible advantages in terms of cost-effectiveness. The feebate is approximately equivalent to a tax on carbon emissions from the power sector, with revenues used to finance a per-unit subsidy for electricity production. More generally if the pivot point is reduced (i.e., the threshold CO_2 per kWh, which determines whether firms pay fees or receive rebates), the feebate has a greater impact on electricity prices (because more generators are paying fees than are receiving subsidies). In this case, the policy is equivalent to an electricity emissions tax, with a fraction of (rather than all) revenues used for a production subsidy.

Energy efficiency policies. Regulatory policies can also reduce the demand for electricity, and direct fuel usage, through setting standards for energy intensity. For example, several countries (e.g., China, Japan, the United States) set standards for the average fuel economy (kilometers per liter or equivalent) of new passenger vehicle fleets. Building codes are also common, as are standards for the energy usage rate of household appliances (e.g., refrigerators), lighting, and heating/cooling equipment. Again, feebates represent a pricing variant of these policies. For example, if applied to passenger vehicles, manufacturers selling relatively fuel-inefficient vehicles would pay a fee in proportion to the difference between the average fuel consumption rate (or CO_2 per kilometer) of their fleet and that for the industry average, multiplied by vehicle sales, while manufacturers with relatively fuel-efficient fleets would receive a corresponding subsidy.

[6] For example, a required share of 20 percent for zero-carbon fuels might be met, say, by a combined share of 10 percent from renewables, hydro, and nuclear and 20 percent from natural gas, if the latter receives half a credit.

- In the power sector, efficiency standards are less effective at reducing emissions than market-based carbon policies. Potentially the most important reason is that efficiency standards do not provide incentives for power generators to reduce CO_2 emissions per kWh. Another reason is that they do not encourage a reduction in the use of energy-using durables and other goods. Furthermore, a range of energy-intensive goods have typically been exempt from regulations (e.g., small appliances, audio and entertainment equipment, assembly lines), yet higher energy prices would provide across-the-board incentives for more efficient versions of these products. And, at least for some transitory period, standards on new products raise their price relative to used products, which can delay the retirement of old (relatively polluting) products. In contrast, higher energy prices will tend to accelerate retirement of older (energy-inefficient) products.

- In the transport sector, efficiency standards are basically identical to CO_2 standards (on a per-kilometer or tonne-kilometer basis) because this sector uses mostly oil-based fuels. These instruments are less efficient than market-based policies. Higher fuel prices provide incentives to reduce vehicle kilometers driven (by raising fuel costs per kilometer) and to buy more fuel-efficient vehicles: Fuel economy standards (or feebates or CO_2 standards) only exploit the latter margin of behavior, which, as a rough rule of thumb, might reduce their effectiveness by about 50 percent relative to a fuel tax.[7]

Regulatory combinations. In short, regulatory policies by themselves provide only limited incentives for reducing CO_2 emissions. However, regulatory (or feebate) combinations, involving a package of measures to reduce the emissions intensity of power generation and to improve the efficiency of major energy-using durables (buildings, vehicles, household appliances), may go a fairly long way in matching the environmental effectiveness of comprehensive, market-based policies. Nonetheless, even under these combination policies, not all emissions reduction opportunities—in particular reduced use of vehicles and other energy-using durables—will be exploited.

The Cost-Effectiveness of Different Policies

A cost-effective policy achieves a given emissions reduction at lowest overall cost to the economy. This matters, not only for its own sake, but also for enhancing prospects that the policy will be sustained over time. To start with, our discussion focuses on costs within the energy sector. These costs are

[7] In fact, by lowering fuel costs per kilometer, the latter policies tend to encourage more vehicle use, although evidence for the United States suggests that this "rebound effect" is relatively modest.

minimized when the cost of the last tonne of emissions reduced is equated across all firms and households. Later, a broader and more appropriate notion of economic cost is considered, which has important implications for the use of revenues from mitigation policies. Box 1.2 provides more discussion of how to think about costs from an economic perspective.

Box 1.2. Understanding the Costs of Emissions Mitigation

The economic, or "welfare," costs of an emissions mitigation policy summarize the costs of all the different, individual actions taken to reduce emissions (leaving environmental benefits aside). These would include, for example, such direct costs as producing electricity with cleaner but more expensive fuels. They also include the less obvious costs to households from driving less, or utilizing fewer energy-using products, than they would otherwise prefer.

It is often easier to define welfare costs by what they are not. They are not measured in terms of *job losses* in industries most directly affected by new policies. Many of those jobs are usually made up by other sectors after a period of adjustment. Welfare costs need not be closely related to changes in gross domestic product (GDP), either. For example, a regulation that leads to the use of a higher priced alternative and raises product prices may actually increase GDP, even though it has positive welfare cost.

Transfers between one segment of society (e.g., consumers) and another (e.g., producers, the government) are not welfare costs. This means that tax revenues raised through carbon taxes themselves are not directly included in welfare costs, nor are outlays on renewable subsidies. As explained below, however, to the extent that new revenue gains/losses imply changes in the rates of broader taxes that distort the economy (e.g., taxes that reduce the return to work effort and capital accumulation), there will be consequences for the overall welfare cost of the policy.

The welfare cost concept has been endorsed by governments around the world for purposes of evaluating regulations, government investments, taxes, and other policies. In the United States, a series of executive orders since the 1970s has required government agencies to perform hundreds of cost-benefit analyses a year, using welfare costs (and welfare benefits) to determine whether their planned "major" regulations are justified from society's perspective.

Source: Authors.

Market-Based versus Regulatory Policies: A First Look

Market-based policies are cost-effective in the sense that all emissions sources covered under the policy are priced at the same rate. Therefore, all firms and households face the same incentives to alter their behavior in ways to reduce emissions up to the point where the cost of the last tonne reduced (e.g., the cost of additional fuel switching in the power sector or the costs to motorists of forgoing trips) equals the price on emissions. For emissions trading systems, cost-effectiveness requires fluid markets, which may not be possible for countries lacking institutions for enforcing property rights or lacking large numbers of market participants.[8]

Market-based policies with and without full coverage of emissions (including, for example, taxes on electricity or individual fuels) are called cost-effective here because they minimize costs within the energy sector *for the emissions reductions that they achieve.* An alternative way of comparing policies is to compare their costs, *for the same effectiveness in terms of reducing emissions.* Under this latter comparison, the market-based policies with partial emissions coverage are not viewed as cost-effective. To achieve the same emissions reduction as under the policy with full coverage, they place too much of the burden on covered sources and none of the burden on other sources, rather than striking the cost-effective balance of reductions across all emissions sources.

Regarding regulatory policies, such as emissions standards and energy efficiency standards, besides their limited effectiveness, they can also perform poorly on cost-effectiveness grounds *if they force all firms to meet the same standard.* For example, it will be relatively costly for a generator heavily dependent on coal to meet a standard for average CO_2 per kWh, compared with a generator that is less dependent on coal. To promote cost-effectiveness, these standards need to be accompanied by extensive credit-trading provisions. These provisions would allow the coal-intensive generator to have higher CO_2 per kWh than the standard by purchasing credits awarded to another generator with CO_2 per kWh lower than the standard. Similarly, under a vehicle fuel economy standard, trading provisions would allow manufacturers or sellers specializing in relatively large vehicles to fall short of the average fuel economy requirement by purchasing credits from a manufacturer specializing in relatively small vehicles for whom exceeding the standard (to obtain credits) is relatively inexpensive. As noted above, credit trading works well only if trading markets are well developed.

However, a more direct way to promote cost-effectiveness, which circumvents the need for any credit trading, is simply to use pricing variants of these policies. For example, under the power sector feebate, coal-intensive generators will opt to pay fees to the government (and exceed the pivot point CO_2 per kWh), while relatively clean generators will receive rebates (for reducing CO_2 per kWh below the pivot point).[9] It is important, however, that

[8] Even well-developed markets can sometimes be subject to manipulation.

[9] In effect, feebates are the tax analogue to emissions or efficiency standards with perfect credit trading.

the tax saved by relatively dirty/energy-inefficient producers from reducing CO_2 by a tonne is the same as the extra subsidy received by relatively clean/energy-efficient producers for reducing CO_2 by a tonne. If not, there will be an excessively costly pattern of emissions reductions across the two types of producers as they face different rewards for reducing emissions.

More generally, for a regulatory combination to be cost-effective, it requires not only credit trading within sectors, but also across sectors, to establish a single price on CO_2 emissions across the economy. Without a uniform price, there is a risk that too much of the burden of emissions reductions will be borne by one sector and too little by another. Similarly, in a feebate package, the implicit price on emissions should be harmonized across sectors.

Box 1.3 discusses some modeling results for the United States that underscore some of the points made so far. It also notes the potential for redundancies when (as is common in practice) governments implement a suite of related policies.

Box 1.3. Modeling Results on the Effectiveness and Cost-Effectiveness of Alternative CO_2 Mitigation Policies

The figure below summarizes a recent study on the projected effectiveness of various policies at reducing domestic, U.S. CO_2 emissions, cumulated over the 2010–2030 period (the height of the bars), and the average welfare costs per tonne reduced, as defined in Box 1.2, over the same period (indicated by the color of the bars). See Krupnick and others (2010), pp. 149–152, for a definition of all the policies. Here we highlight just a few points.

Not surprisingly, comprehensive carbon taxes and cap-and-trade systems of the scale envisioned in (unsuccessful) federal cap-and-trade bills (and indicated by the set of blue, relatively tall bars) are found to be the most effective at reducing domestic emissions. The average costs reduced are also relatively low for these policies ($11 to $12 per tonne of CO_2 reduced, in 2007 U.S. dollars). Combining a cap-and-trade policy with a renewable portfolio standard (RPS) has essentially no effect on emissions (i.e., the RPS is redundant) as emissions are fixed by a series of annual caps. If domestic sources must meet the same caps, but without any purchases of emissions offsets (offsets are defined below), the domestic emission reduction is larger, though average costs per tonne rise (the extreme left-hand bar).

Emissions reductions under the RPS by itself are only about 25 percent of those under broad pricing policies. But allowing credits for incremental natural gas (RINGPS) or credits for all fuels with lower carbon intensity than coal—the Clean Energy Portfolio Standard (CEPS-ALL)—substantially improves effectiveness of up to about 50 to 60 percent of that under broader emissions pricing policies.

Box 1.3. (*continued*)

However, even very large increases in gasoline taxes (of about US$1 per gallon or US$0.26 per liter) reduce emissions by only a minor fraction of the reduction under broad pricing policies. Most obviously, this policy only covers emissions from road transport. In addition, options for substituting clean fuels for conventional fossil fuels in passenger vehicles are limited (compared with fuel switching possibilities in the power sector). And manufacturers are already incorporating advanced fuel-saving technologies to meet escalating Corporate Average Fuel Economy (CAFE) standards.

Another policy redundancy—in the presence of binding CAFE requirements—is subsidies for hybrid vehicles. These subsidies lead to a greater penetration of hybrids, but manufacturers can then ease up on improvements for conventional gasoline vehicles while still meeting the same fleet-wide average fuel economy standard.

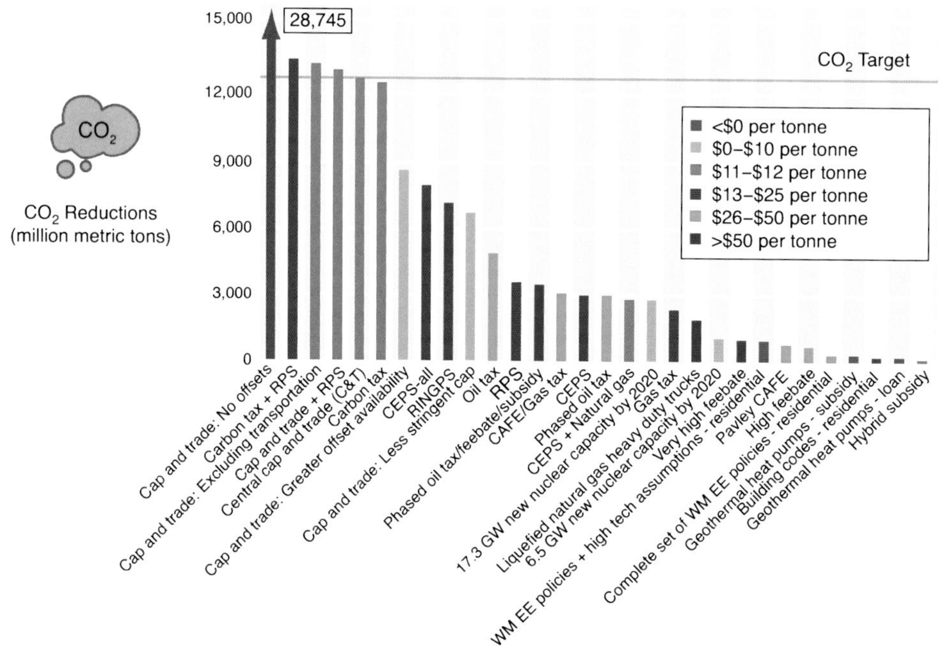

Source: Authors, selected cases from Krupnick and others (2010) based on simulating a variant of the U.S. Energy Information Agency's National Energy Modeling System.

A Closer Look at Cost-Effectiveness

Comprehensive carbon taxes, as well as cap-and-trade systems with allowance auctions, provide a potentially significant source of annual government revenue—perhaps in the order of 1 percent of GDP for the United States and over 2 percent for China. How this revenue is used will have important implications for the broader costs of market-based policies beyond the costs in energy markets.

In particular, if these revenue sources are used to reduce other taxes that distort the broader economy, then this can help to substantially reduce overall policy costs. Taxes on labor income, for example, distort the labor market by lowering the returns to labor force participation and effort. Taxes on corporate income and income from household savings distort the capital market by reducing capital accumulation below levels that would otherwise maximize economic efficiency. Using climate policy revenues to cut these taxes therefore produces broader benefits to the economy.

Despite these potential benefits, the overall costs of carbon taxes, as well as cap-and-trade systems with allowance auctions, are likely to be positive (although, up to a point, environmental benefits will be much larger than these costs). This is because there is a counteracting effect that offsets the benefits from revenue recycling—as carbon taxes and cap-and-trade systems drive up energy prices, they tend to contract (albeit very slightly) the overall level of economic activity, which in turn has a (slightly) depressing effect on employment and investment.

The main point here (as discussed further in Chapter 2) is that how revenues are used can have important implications for the overall costs of market-based instruments. If revenues from carbon taxes are used in socially productive ways, such as to reduce distortionary taxes elsewhere in the economy or fund socially desirable spending, then this substantially lowers policy costs. Similarly, for cap-and-trade systems to be cost-effective, allowances need to be auctioned and revenues need to be used productively. If instead allowances are given away for free in a lump-sum fashion to industry, overall (net) policy costs are substantially higher as a valuable revenue-recycling benefit is given up. In fact, allocating all the allowances free to affected industries will greatly overcompensate them, given that most of the allowance price tends to be paid by households in the form of higher energy prices rather than paid by firms in the form of lower producer prices.

If revenues from taxes or cap-and-trade are not used wisely, certain regulatory combinations may conceivably perform better on overall cost-effectiveness grounds than market-based policies. In this regard, a "benefit" of regulatory instruments is that they tend to have a weaker effect on energy prices than market-based policies because they do not involve the pass-through of tax revenues or allowance rents in higher prices. Consequently, regulatory policies can do less harm to overall economic activity than market-based approaches that do not exploit the revenue-recycling benefit. For most countries, the best policy of all on cost-effectiveness grounds is a carbon tax, or auctioned cap-and-trade system, with revenues used to cut broader distortionary taxes, either directly or indirectly through deficit reduction (which avoids the need for raising other taxes).

Dealing with Uncertainty

The future costs of emissions control instruments will also depend on the future prices of clean and dirty fuels and the future cost of emissions-saving technologies. Considerable uncertainty surrounds these factors. Given the strong desire of environmental groups and others to fix the quantities of emissions (or renewables), such groups tend to favor a cap-and-trade system (and quantity mandates) rather than a fixed price system (e.g., a tax), as the latter lets quantities vary over time as uncertainties are resolved.[10] Yet, in the presence of uncertainty, there is a cost to fixing the emissions limit (for covered sources) involving (1) allowance price volatility that causes too little abatement in some years and too much in others from a cost-effectiveness viewpoint and (2) the slowing of long-term, clean technology investments. There are ways to deal with these concerns, but only if policymakers are willing to relax rigid annual emissions controls.

Taxes versus Cap-and-Trade

Annual emissions targets leave the price of allowances in cap-and-trade systems to be determined by the market. Prices are relatively high in periods when meeting the cap is costly (e.g., in times of high energy demand or high prices for clean fuels) and vice versa in periods when the costs of meeting the cap are relatively moderate. Reducing price volatility can help to lower program costs over time for a given cumulative reduction in emissions. With a stable emissions price (or rather, one rising at the interest rate), emissions reductions will be greater in periods when the costs of those reductions are relatively low and vice versa when controlling emissions is relatively costly: in this way, stable prices help to equate the (discounted) costs of incremental abatement in different years. Stable emissions prices may also create business conditions that are more conducive to investments in clean technologies (e.g., wind and solar plants) with high upfront costs and long-range payoffs in terms of emissions reductions.

One way to limit price volatility in a cap-and-trade system is to allow firms to bank allowances (i.e., carry forward allowances to cover emissions in future

[10] Environmental groups have an aspiration for environmental certainty, implying a preference for quantity over price/cost targets. However, unless a cap-and-trade policy covers all sources of CO_2, such certainty cannot be attained. And even if a given country fully covers its sources with such a program, carbon leakage to countries without a policy will create quantity uncertainty.

years rather than turning them all in now), which enables them to do extra abatement in periods when emissions reduction costs are low. Another is to allow advance auctions where firms can buy permits at today's prices for use several years from now (if they anticipate higher permit prices). Furthermore, firms might borrow allowances (i.e., use some allowances for future periods now), which enables them to do less abatement when emissions control costs are high.

Another possibility is to combine a cap-and-trade system with a price collar. In periods when allowance prices hit a ceiling level, the government could sell extra allowances to the market at that ceiling price, thereby relaxing the emissions cap, while in periods when allowance prices fall to a floor level, the government could step in and buy allowances back at the floor price, thereby tightening the emissions cap.

Yet another possibility is to allow covered sources to purchase international emission offsets (e.g., through the Clean Development Mechanism), which helps to put a ceiling on the domestic allowance price. Offset provisions enable domestic firms to claim credits by paying for (cheaper) mitigation projects, typically in developing economies. Offsets are not always real, however (i.e., the developing economy project may have occurred anyway without the offset payment), in which case environmental effectiveness is undermined (and the domestic country makes a transfer to the developing economy for no emissions benefit). Preserving policy credibility may therefore require stringent verification requirements for offsets, implying a correspondingly higher emissions price and domestic abatement cost.

However, the best way to provide price stability is simply to implement a carbon tax (with the price rising automatically at a fixed annual rate) instead of a cap-and-trade system. The tax provides full (rather than partial) price stability, without the need for complicating design provisions.

The drawback of price stability is that policymakers lose control of annual CO_2 emissions from covered sources—annual targets for specific years have so far underpinned international negotiations over climate mitigation. However, one year's emissions by one country have essentially no impact on future global warming—rather, this is determined by the historical accumulation of global emissions since the industrial era. If policymakers continue to negotiate over quantities rather than emissions pricing, a better approach than annual targets might be to focus on carbon budgets. These budgets would fix allowable cumulative emissions over a multiyear period (say, 10 years), leaving countries with flexibility over annual emissions.

Other Policies

Following similar logic, price-based alternatives to regulatory policies are better able to handle uncertainty over future abatement costs, although at the (political) expense of variability in year-to-year emissions. For example, a feebate for the power sector (with the emissions price growing at the rate of interest) will equate the (present value) of the incremental cost of abatement in different years. A strict CO_2 per kWh standard each year would not be cost-effective, as the incremental costs of meeting the same standard are likely to vary over time as fuel prices, and so forth, change. Again, this problem could be addressed, at least in part, through price stability provisions (banking and borrowing of credits, price ceilings, floors).

Incidence and Competitiveness

The burden of climate policies on households (especially poor households), firms, and the implications for the competitiveness of industries producing tradable products are often major concerns to policymakers. These burdens stem from the effect of policies on energy prices, particularly electricity prices, but also on fuels directly consumed by households and firms. Chapter 2 discusses these issues in the context of carbon taxes, along with possibilities for offsetting household and industry burdens. Here we simply compare the seriousness of distributional and competitiveness effects of other instruments relative to those for carbon taxes.

Burden on Households

In developed economies, poorer households tend to spend a relatively large portion of their income on electricity, transportation fuels, and fuels for heating and cooking. This means that the burden—relative to income—of the higher energy prices (caused by comprehensive carbon pricing policies) is greater for lower income households, which runs counter to broader government efforts to moderate income inequality. For developing economies, the burden-to-income ratio might be lower for relatively low-income groups if they do not own vehicles or have access to electricity. Nonetheless, any new policy that potentially reduces living standards in absolute terms for the poor may require offsetting compensation.

Clearly, the burden on low-income households will be less severe for market-based instruments with partial coverage or for individual taxes on electricity or vehicles, but these policies have very limited environmental effects. More important is the distinction between market-based policies and regulatory combination policies, or feebate combinations, with broad environmental effectiveness. As already mentioned, market-based policies can have a much bigger effect on energy prices, as they involve the pass-through of large revenues from taxation or permit auctions or of allowance rents (if not auctioned) into higher prices.

Burden on Firms

Any policy that raises the price of products—which includes most policies to reduce carbon emissions—will have effects across sectors that compete with one another (such as coal versus natural gas sales to electricity producers) and/or compete with countries that do not apply similar charges. The industries hit hardest are energy-intensive, trade-exposed sectors, where there are limits on the pass-through of input costs to product prices. For example, higher electricity prices will hurt those industries that are heavy electricity users, like aluminum producers and oil refiners. Aside from the political problems posed by firms that fear being outcompeted, there are concerns about job outsourcing and carbon leakage.[11]

Implications for Instrument Choice

In fact, distributional incidence may provide a second reason for revisiting the case for market-based instruments over regulatory and feebate approaches (the first reason being the possibility that the actual or potential revenue recycling benefits from pricing instruments are not exploited). If households and industry cannot be adequately compensated under market-based policies, it may well be that the practical benefits of avoiding large increases in energy prices through using other instruments outweigh the drawbacks of those instruments (in terms of missing some emissions reduction opportunities).

Naturally, there are caveats here. As noted, regulatory and feebate approaches would need to be comprehensive and harmonized to provide the same rewards for additional emissions reductions across different sectors. Moreover, at more stringent levels of abatement, as opposed to moderate abatement levels, the relative discrepancy in energy price impacts between market-based and other approaches becomes less pronounced.[12] That is, the practical advantages of other instruments diminish as the policy is tightened over time. Even if, for example, feebates were the preferred instrument

[11] While difficult to project accurately, the problem of this source of emissions leakage should not be overstated. For example (leaving aside well-integrated regions like the European Union), reductions in transportation fuels in one country or shifts to cleaner power-generation fuels are likely to cause little offsetting increases in emissions in other countries (at least in the absence of significant reductions in world fossil fuel prices).

[12] For example, at higher tax levels, emissions per kWh are lower, implying a smaller impact on electricity prices from further tax increases.

initially, ideally there would be a progressive transition to market-based instruments as the feasibility of the latter improves.

Nonetheless, the ideal approach would be to start with a market-based instrument but provide the needed compensation to adversely affected groups—so long as this compensation does not compromise policy costs too much. As discussed in Chapter 2, there are some promising ways to do this.

Promoting Clean Technology Development and Deployment

In this chapter, we have examined alternative instruments to correct for the market failure of uninternalized externalities associated with CO_2 emissions. Such instruments, particularly carbon taxes or a cap-and-trade approach, also stimulate the creation and deployment of new technologies—any new way of reducing CO_2 emissions at a cheaper cost will be of interest to emitters if the cost of acquiring and using that technology is less than their outlays for the CO_2 emissions such technologies would displace. Broad-based pricing instruments provide incentives for clean technology development and deployment across all sectors of the economy.

However, uncertainty over future emissions prices—as in cap-and-trade systems lacking price stability provisions or carbon taxes where future tax rates are not well defined—may deter clean technology investments. Moreover, if the tax or cap-and-trade system has partial, rather than full, coverage, it will lack the across-the-board technology incentives provided by more comprehensive pricing. Similarly, taxes on electricity or individual fuels incentivize only a narrow range of clean technology investments.

Feebates or emissions standards are superior to specific technology standards (e.g., CCS) because, for the latter, once the targeted technology is adopted, the incentive to develop new technologies stops.[13] But again, the former needs to be implemented and coordinated across sectors to provide the broader technology incentives that are automatic under comprehensive carbon pricing policies.

Even with CO_2 emissions comprehensively priced, there are reasons for believing that efforts to invent, develop, and deploy new clean technologies will be inadequate because of additional market failures. In general, this calls for use of supplementary and targeted technology policies, rather than setting emissions control instruments more aggressively. Box 1.4 provides some discussion of the rationale for and type of technology policy.

[13] In fact, after the race to establish technology standards is over, the regulated community may actively move away from developing better technologies for fear of opening up new rule-making.

Box 1.4. The Potential Case for Complementary Technology Policies

Generally, economists recommend that technology-related market failures associated with basic research, applied research and development (R&D) at firms, and technology deployment require their own instruments. There are some general caveats to bear in mind, however:

- *Technology policies should be a complement to, not a substitute for, emissions mitigation policies.* As noted above, emissions pricing is the single most effective policy to reduce emissions (given current technology) and also stimulate clean technology investments.

- *In general, the playing field should not be tilted in favor of one specific technology over others.* So policies to subsidize carbon capture and storage or that mandate use of certain types of alternative-fuel vehicles rather than stimulating all comers could be inefficient unless the market failures are especially severe for the favored technologies.

- *Innovative activity in the public sector or the energy sector may "crowd out" such activity elsewhere in the economy.* For example, new scientists and engineers working on energy technologies might have previously worked in other sectors.

These factors suggest that technology policies need to be carefully scaled and designed. Which instrument is appropriate and how long it should be applied depend on the nature of the market failure. There are several possibilities for technology-related market failures, though some are less convincing than others.

There is a potentially strong case for policies encouraging basic research in publicly funded institutions and applied R&D at firms. In particular, the "public goods" problem—that is, the inability of innovators to capture spillover benefits to other potential users from technology breakthroughs—is most severe at this stage of the innovation process. Indeed, for the United States, numerous studies show that the social rate of return to basic R&D (i.e., including benefits to all potential users) is several times the private rate of return.[14] Although the problem applies to innovation in general, it can be more pronounced for clean energy technologies, given that many of them (e.g., renewable plants) have high upfront costs and long-range payoffs and that there is uncertainty regarding future governments' commitments to emissions pricing.

What Market Failures Might Justify Additional Support for Energy-Related Technologies?

Early producers of new technologies often invoke the "infant industry" argument that a fledgling sector needs protection from world markets, say through tariffs or nontariff

[14] Likewise, policies that encourage general education and training of innovators are desirable because any one employer who engages in such activities may see its employee leave for another job.

Box 1.4. (*continued*)

trade barriers. But this argument means little for economic efficiency in the country as a whole (and in the short term will reduce economic efficiency) and, if accepted, requires a strict criterion for judging when the industry has "grown up."

A potentially more solid case for technology policies arises if firms are reluctant to adopt new technologies because they would bear all the costs of "learning by doing," which benefits later users of the technology. This provides a possible rationale for clean technology deployment policies. But policies should be transitory and phased out as the technology matures. Moreover, gauging the future penetration rate of a new technology can be difficult given uncertainty over its costs and that of competing technologies, suggesting the desirability of a flexible pricing instrument (e.g., a subsidy) over a quantity instrument (e.g., a minimum sales share requirement for electric vehicles) that forces the new technology regardless of its costs. And there is a danger of creating an uneven playing field if some technologies are favored at the expense of others.

Another argument for technology deployment policies is that consumers' demand for energy-efficient investments is held back by their myopia—they seem unwilling to make a big investment today that will pay for itself in several years, rather than over the entire lifetime of the investment. For this argument to stand, we need to distinguish between "hidden" costs and market failure. If consumers are reluctant to buy because the technologies are unproven or the costs are hidden (e.g., reluctance to buy compact fluorescent lights reflects their perceived lower quality compared with incandescent light bulbs), this is not a justification for intervention.

On the other hand, consumers may lack information about the features and lifetime energy savings of particular technologies. Alternatively, the person making the purchase decision (e.g., a landlord) may not care about energy savings if these benefit someone else (a tenant responsible for paying energy bills). Furthermore, capital markets may unreasonably deny households access to credit to make large investment purchases. In principle, these market failures would justify some form of policy intervention such as information campaigns if the problem lies in that area, reform of tenant-landlord interactions, measures to increase credit availability, or incentives for clean technology adoption.

Finally, policies such as subsidies or prices that target the improvement of networks (e.g., new pipeline infrastructure for clean fuels) are also potentially warranted. In these cases, the benefits of the technologies to other firms may be so pervasive that no single private company can appropriate them all. Alternatively, the risk of the technology failing may be higher than a private concern can handle but may be acceptable to a government, which has more opportunities to hedge against such failure and has lower costs of accessing funds.

Source: Authors.

Conclusion

The choice of instruments to reduce CO_2 is a complex one. In this chapter, we have laid out the basics of a comparison of instruments according to five criteria, and the main points are summarized in matrix form in Table 1.1.

Market-based instruments are potentially the most effective policies for reducing emissions, although raising revenue and using that revenue productively are important for containing their overall policy costs. The choice between carbon taxes and cap-and-trade systems is less important than implementing one of them and getting the design details right, which include comprehensive coverage of emissions, exploiting the fiscal dividend, and (in trading systems) limiting price variability, although only carbon taxes may be viable if institutions for credit trading are lacking. If carbon pricing policies are not initially acceptable, a combination of regulatory policies can be a reasonable alternative for the time being if they are carefully chosen to mimic, insofar as possible, the emissions reduction opportunities that would be exploited under comprehensive pricing policies and they include extensive credit trading provisions. In the latter regard, using feebate alternatives to regulations is simpler as it avoids the need for institutions to enforce credit trading.

Table 1.1. Summary Comparison of Policy Instruments

Policy Instrument	Effectiveness at Reducing Economy-Wide CO_2	Cost-Effectiveness[a]	Dealing with Uncertainty over Abatement Costs[b]	Promoting Clean Technology Deployment	Incidence and Competitiveness	Overall Assessment
Comprehensive carbon taxes (upstream)	Most effective policy	Cost-effective[c]	Automatically accommodates uncertainty	Effective, though supplementary measures to overcome technology barriers may be needed	Energy price impact can burden low-income households and harm competitiveness	Potentially the best policy, but incidence and competitiveness effects may need addressing
Comprehensive cap-and-trade (upstream)	Same as comprehensive carbon tax	Cost-effective if allowances auctioned[c]	Price stability provisions needed	Same as comprehensive carbon tax (with price stability provisions)	Same as comprehensive carbon tax if allowances are auctioned (but incidence can change if allowances are freely allocated)	Same as comprehensive carbon tax (1) if allowances are auctioned, (2) there are price stability provisions, (3) there are well-functioning credit markets
Carbon tax with partial coverage (downstream)	Partially effective	Cost-effective[c]	Automatically accommodates uncertainty	Promotes narrower range of technology investments	Similar issues as under comprehensive carbon tax	Potentially attractive initially (in absence of comprehensive tax)
Cap-and-trade with partial coverage (downsteam)	Same as partial carbon tax	Cost-effective if allowances auctioned[c]	Price stability provisions needed	Same as partial carbon tax (with price stability provisions)	Same as partial carbon tax if allowances are auctioned (but incidence can change if allowances are freely allocated)	Same as partial carbon tax (1) if allowances are auctioned, (2) there are price stability provisions, (3) there are well-functioning credit markets
Pure electricity tax	Limited effectiveness	Cost-effective for small emissions reductions[c]	Automatically accommodates uncertainty	Promotes a very narrow range of clean technologies	Similar issues as under the comprehensive carbon tax	Generally not recommended (unless combined with other mitigation instruments)
Simple excise tax on vehicle purchases	Very ineffective	Cost-effective for very small emissions reductions[c]	Uncertainty is not an issue	There is essentially no effect	Imposes burden on motorists	Not recommended on environmental grounds

Taxes on individual fuels	Limited, though some taxes (on coal) are more effective than others (on gasoline)	Cost-effective for modest emissions reductions[c]	Automatically accommodates uncertainty	Promotes limited range of clean technologies	Some burden on households and firms	Inferior to comprehensive emissions pricing
Incentives for clean generation fuels	Limited effectiveness	Fairly cost-effective (for modest emissions reduction) if there are credit trading provisions for quantity instruments	Price instruments accommodate uncertainty, quantity instruments require price stability provisions	Promotes limited range of clean technologies	Fairly small burden on households and firms (for moderately scaled policy)	Inferior to comprehensive emissions pricing
Emissions standards (for power sector)	Fairly effective (for power sector)	Cost-effective if credit trading provisions	Price stability provisions needed	Provides little incentives for electricity-saving technologies or technologies in other sectors	Fairly small burden on households and firms (for moderately scaled policy)	Promising if comprehensive market-based policy is not feasible, but it should be combined with other policies
Energy efficiency standards	Limited effectiveness	Cost-effective (for modest emissions reduction) if credit trading across firms	Price stability provisions needed	Promotes only limited range of technology investments	Relatively modest burden on households and firms	Not a substitute for emissions pricing, but could play a useful role in regulatory combination
Feebates	Fairly effective (for power sector)	Cost-effective (for modest to partial emissions reductions)	Automatically accommodates uncertainty	Promotes some technology investments	Modest burden on households and firms	Promising in absence of comprehensive emissions pricing, but several schemes required for different sectors
Regulatory combination[d]	Potentially fairly effective	Fairly cost-effective if there is credit trading across firms and sectors	Price stability provisions needed	Promotes a fairly broad range of technology investments	Fairly modest burden on households and firms for moderately scaled policy	Promising in absence of comprehensive emissions pricing, if there is extensive credit trading across sectors

Source: Authors.

[a] Compares costs for the *different* level of emissions reductions achieved by different policies.

[b] Note the limited treatment of uncertainty in this column.

[c] Assumes revenues are used productively to improve economic efficiency, such as to reduce other distortionary taxes.

[d] Combining energy-efficiency standards for major products (e.g., vehicles, buildings, household appliances) with emissions standards for power generation.

References and Suggested Readings

For a general discussion comparing a broad range of alternative carbon mitigation instruments, see the following:

Aldy, Joseph E., and Robert N. Stavins, 2011, "Using the Market to Address Climate Change: Insights from Theory and Experience," Discussion paper RWP11-038 (Cambridge, Massachusetts: Harvard University Kennedy School of Government).

Goulder, Lawrence H., and Ian W. H. Parry, 2008, "Instrument Choice in Environmental Policy," *Review of Environmental Economics and Policy,* Vol. 2, pp. 152–174.

Krupnick, Alan J., Ian W. H. Parry, Margaret Walls, Tony Knowles, and Kristin Hayes, 2010, *Toward a New National Energy Policy: Assessing the Options* (Washington: Resources for the Future and National Energy Policy Institute).

General issues in the choice between carbon taxes and emissions trading systems are covered in the following:

Hepburn, Cameron, 2006, "Regulating by Prices, Quantities or Both: An Update and an Overview," *Oxford Review of Economic Policy,* Vol. 22, pp. 226–247.

Nordhaus William, 2007, "To Tax or Not to Tax: Alternative Approaches to Slowing Global Warming," *Review of Environmental Economics and Policy,* Vol. 1, pp. 26–44.

For a focus on the importance of revenue recycling for containing the costs of market-based policies, see the following:

Parry, Ian W. H., and Roberton C. Williams, 2012, "Moving US Climate Policy Forward: Are Carbon Tax Shifts the Only Good Alternative?" in *Climate Change and Common Sense: Essays in Honor of Tom Schelling,* ed. by Robert Hahn and Alistair Ulph (Oxford, UK: Oxford University Press, pp. 173–202).

For a discussion of possible manipulation in allowance trading markets, see the following:

Stocking, Andrew, 2010, "Unintended Consequences of Price Controls: An Application to Allowance Markets," Working Paper 2010–06 (Washington: Congressional Budget Office), September.

For some discussion of instruments for promoting clean fuels in power generation, see the following:

Aldy, Joseph E., 2011, "Promoting Clean Energy in the American Power Sector," Hamilton Project discussion paper 2011–04 (Washington: Brookings Institution).

Palmer, Karen, Richard Sweeney, and Maura Allaire, 2010, "Modeling Policies To Promote Renewable and Low-Carbon Sources of Electricity," Background paper for Krupnick, Alan J., Ian W. H. Parry, Margaret Walls, Tony Knowles, and Kristin Hayes, 2010, *Toward a New National Energy Policy: Assessing the Options* (Washington: Resources for the Future and National Energy Policy Institute).

Chapter

2 How to Design a Carbon Tax

Ian Parry
Fiscal Affairs Department, International Monetary Fund

Rick van der Ploeg
University of Oxford, United Kingdom

Roberton Williams
University of Maryland and Resources for the Future, United States*

Key Messages for Policymakers

- Market-based instruments like carbon taxes are potentially the most effective policies for reducing energy-related CO_2 emissions. They do this by cutting the demand for fossil fuels and making it more attractive to use zero-carbon fuels like renewables.
- Ideally, taxes should be applied where fossil fuels enter the economy with rates levied in proportion to carbon content and refunds for any downstream carbon sequestration.
- To keep down overall policy costs, carbon tax revenues should be used to alleviate distortions created by the broader fiscal system, reduce government debt, and/or fund valuable government expenditures. Revenues might also fund climate adaptation (e.g., water defenses) where private sector investment would otherwise be too low. Care should be taken not to use valuable revenue for inefficient subsidies (e.g., fuel subsidies).
- Cutting preexisting, environmentally blunt energy taxes (e.g., excises on electricity or on vehicle ownership) may help to compensate adversely affected groups for higher energy prices, thereby enhancing feasibility. Alternatively, low-income groups and firms in trade-exposed sectors might be compensated through targeted measures such as adjustments to the broader tax/benefit system and transitory production subsidies.

* This chapter has benefited from the constructive comments of Joseph Aldy, Terry Dinan, Daniel Hall, Michael Keen, Richard Morgenstern, and Vicki Perry.

- Non-CO_2 greenhouse gases (GHGs) might be covered directly under the tax, or indirectly through emissions offset credits, as capability for monitoring and verification is developed over time.
- Carbon pricing policies are the most important instruments for promoting the development and deployment of clean technologies. Supplementary instruments may be needed to help overcome market barriers to large clean-energy investments, though they need to be carefully designed.
- At an international level, a carbon tax floor negotiated among major emitters is a potentially promising way forward (and could be easier to negotiate than multiple country-level emissions targets).

Carbon pricing policies are potentially the best instruments for incorporating environmental damages into the market price of intermediate inputs and the price of final goods and services. By reducing the demand for fossil fuels and discouraging exploration for new fossil fuel reserves—especially fuels with high carbon dioxide content, these pricing effects exploit the entire range of behavioral changes at both the household and firm levels for reducing energy-related CO_2 emissions. Carbon pricing also creates across-the-board incentives for the development and deployment of clean-energy technologies, and it boosts the supply of carbon-free renewable substitutes. A carefully designed time path for carbon pricing not only ensures that firms and households use less CO_2-intensive production methods, appliances, vehicles, machines, and so forth, but also ensures that renewables are brought to the market and phased in more quickly. Especially attractive in the present fiscal crisis, carbon pricing can also provide a substantial source of government revenue. Here the focus is on carbon taxes, though as explained in Chapter 1, (revenue-raising) emissions trading systems are also very promising approaches.

In crafting carbon tax legislation, policymakers may be concerned about a number of design issues, such as the following:

- The choice of the tax base
- How tax revenues might be used
- What might be done to address concerns about distributional effects and competiveness
- How to simplify administration and compliance
- To what extent non-energy-related emissions and emission offsets might be integrated

- Whether supplementary instruments are needed to promote clean technology investments
- How, at an international level, a carbon tax agreement might be negotiated and monitored

This chapter discusses these issues in turn. Other important questions, such as the appropriate level and time path of carbon taxes, the pros and cons of taxes versus other emissions control instruments, appropriate policies for low-income countries, and complementary reform of energy subsidies, are discussed elsewhere in this volume. At the end of this chapter, we briefly evaluate some existing tax systems in light of our recommendations.

Choice of Tax Base

The most important consideration in choosing the base of a carbon tax is to maximize the coverage of emissions sources (thereby avoiding implicit or explicit exemptions to significantly polluting activities), although beyond some point, further extensions to the tax base may not justify the additional administrative and compliance complexities. All potential CO_2 emissions across different fuel types and fuel users should be taxed at the same rate, as they all cause the same environmental damage regardless of how they are generated or in which location.[1]

Ideally, carbon taxes should be levied upstream in the fuel supply chain to maximize coverage, while also limiting (for administrative reasons) the number of collection points (see later). And charges should be levied in proportion to the fuel's carbon content to equate prices across (potential) emissions releases.[2] Refunds should be provided to encourage downstream carbon capture and storage (e.g., at coal-fired power plants) if and when these technologies become viable.[3]

Uniform charging for carbon alters absolute and relative energy prices, providing firms and households with incentives to exploit all of the major opportunities for reducing fossil fuel CO_2 emissions. These opportunities include reducing

[1] There could be special cases where exemptions are warranted on economic grounds, although these cases need to be carefully evaluated. One possibility is that excise taxes on transportation fuels are already excessive (prior to further tax increases from carbon charges) in that they may "overcorrect" for other problems like road congestion, accidents, and local tailpipe emissions, although these problems typically warrant fairly high fuel taxes (see, e.g., Parry, Walls, and Harrington, 2007).

[2] For example, combusting a liter of fuel oil or diesel produces about 0.0027 tonnes of CO_2, a liter of gasoline produces about 0.0023 tonnes, and combusting a (short) tonne of coal produces about 2.45 tonnes of CO_2 (see http://bioenergy.ornl.gov/papers/misc/energy_conv.html). To convert CO_2 to carbon emissions, divide by 3.67.

[3] The refund should equal the CO_2 tax times the quantity of CO_2 that is captured and permanently stored. The tax paid on the carbon content of fossil fuel acts like a deposit that is then refunded if the carbon is captured and stored.

electricity demand, improving power plant efficiency (i.e., reducing fuel inputs required per kilowatt hour [kWh] of generation), and shifting from generation fuels with high carbon intensity (coal) to fuels with intermediate carbon intensity (fuel oil, natural gas) and from these fuels to zero carbon fuels (nuclear, hydro, other renewables). They also include reducing the demand for electricity, transportation fuels, and direct fuel use in homes and industry through improving the efficiency of energy-using products (e.g., vehicles, lighting, household appliances) and reducing the use of these products (e.g., reducing kilometers driven per vehicle, economizing on the use of air conditioners). A uniform price on CO_2 provides the same incentive at the margin to reduce emissions across all these possibilities, since everybody gets the same savings for altering behavior in ways that reduce emissions by an extra tonne.

Table 2.1. Percent Increase in Energy Prices from a US$22 Per Tonne of CO_2 Tax, Selected Countries, 2009

Fuel	Steam coal	Diesel		Electricity		Light Fuel Oil		Natural Gas		Gasoline (regular unleaded)
End user	Generators	Industry	Households	Industry	Households	Households	Generators	Industry	Households	Households
Canada	200.0	7.6	na	8.3	5.8	8.7	na	27.1	11.8	6.2
Taiwan	57.7	8.2	8.2	18.6	15.8	na	10.0	8.9	9.2	6.5
France	44.7	5.1	4.2	1.7	1.1	7.4	na	10.7	5.5	na
Germany	46.2	4.6	3.9	8.5	3.7	8.1	na	8.4	4.1	2.9
Indonesia	72.2	9.3	12.8	25.6	28.0	24.5	35.3	na	na	na
Italy	49.4	4.7	3.9	4.2	4.1	4.1	na	8.4	4.4	na
Japan	na	6.8	5.3	5.8	4.0	8.3	na	8.3	3.0	4.0
Mexico	99.5	11.8	10.3	15.2	16.4	na	23.6	na	11.1	9.2
Netherlands	na	5.1	4.3	7.5	4.1	6.4	na	9.3	4.0	na
Poland	63.3	6.2	5.1	13.4	9.6	7.4	16.7	10.8	5.8	na
Republic of Korea	61.6	na	5.4	16.7	12.6	7.8	10.3	9.8	8.1	4.2
Spain	na	5.4	4.7	9.4	4.6	7.7	na	10.8	5.1	na
Thailand	na	8.2	na	16.8	13.0	5.4	na	16.3	na	5.7
Turkey	158.4	3.5	3.5	9.3	7.8	4.0	10.0	10.0	8.2	na
United Kingdom	60.0	4.2	3.6	7.8	5.1	8.6	18.4	14.5	5.8	na
United States	100.8	9.1	9.1	21.8	12.9	8.4	24.9	22.9	10.0	8.4

Sources: Carbon coefficients for fossil fuels are from www.eia.gov/oiaf/1605/coefficients.html; energy prices are from *Energy Prices and Taxes Statistics* accessed through the OECD ILibrary; and CO_2 per kWh (averaged from 1992 to 2002) is from http://205.254.135.7/oiaf/1605/pdf/Appendix%20F_r071023.pdf.

Note: The absolute price increase for each fossil fuel is given by its CO_2 coefficient per unit (obtained from U.S. data) times US$22 per tonne. The absolute increase in electricity prices is calculated by the average CO_2 per kWh for generation in a country, times the CO_2 tax. These absolute price increases are compared with prevailing prices in 2009 across different countries to obtain the percent price increase. It is assumed that the tax is fully passed forward—in reality, some of the tax may be passed backward in the form of lower prices received by fuel suppliers. na = not available.

Table 2.1 illustrates (retrospectively for 2009) the potential impact of a US$22 per tonne (€16) CO_2 charge on various energy prices in selected countries, assuming full pass-through of the tax into prices (which should be a reasonable approximation over the longer term for individual countries).[4] In proportional terms, the greatest impact is on coal prices, which rise by 45 to 200 percent across the countries. Retail gasoline prices, in contrast, rise by 3 to 9 percent, and residential natural gas prices rise by 3 to 12 percent—gas prices for generators and industry tend to rise more in relative terms as they pay lower prices (e.g., due to bulk buying). The estimated impact on residential electricity prices varies between 1 percent (in France where generation is largely nuclear) and 28 percent (in Indonesia, where electricity is heavily subsidized)—again, relative price increases are larger for industry.

Other (less-efficient) tax bases include downstream systems that tax emissions released from major stationary sources (e.g., coal and natural gas plants, metal manufacturers). These may be a more natural extension of earlier local pollution programs, and they encompass the lowest cost abatement opportunities (which are usually in the power sector).

However, downstream programs tend to exempt entities with emissions below a certain threshold, and they need to be accompanied by additional programs to cover transportation and home heating fuels. In the European Union, the Emissions Trading System (ETS) has a downstream focus and misses out on about 50 percent of energy-related CO_2 emissions. Although an EU-wide carbon tax on fuels currently exempt from the ETS has been proposed, better still would be one upstream program applying a uniform carbon price to all fossil fuels.[5]

Excise taxes on electricity use are common among the Organization for Economic Cooperation and Development (OECD) countries (though they are more significant at the residential than industrial level). They are often justified on climate grounds, but they are far less effective at exploiting emission reduction opportunities than comprehensive carbon taxes. Electricity taxes provide no reward for switching to cleaner generation fuels, improving power plant efficiency, or reducing emissions outside the power sector. Within the power sector, coal taxes are somewhat better, as they encourage shifting

[4] For fossil fuels, the absolute price increase—the CO_2 content of the fuel in tonnes times US$22—is taken to be uniform across countries, while the percent price increase varies due to differences in prior fuel prices across countries. For electricity, the absolute price increase also varies across countries depending on their generation mix, which determines CO_2 per kWh.

[5] It is sometimes suggested that a downstream tax will have greater effect as it is more visible than an upstream tax. However, in an upstream system, firms are likely to pass forward the embedded carbon tax in higher prices for coal, gasoline, and other fuels, so power generators, motorists, and so on would be fully aware of the fuel price increase.

away from high-carbon-generation fuels, though they still fail to encourage shifting from natural gas and fuel oil to zero-carbon fuels.[6]

Vehicle ownership taxes (e.g., excise taxes, registration fees, annual road taxes) are especially common. Even within the transportation sector, however, these taxes do not (in general) encourage households with vehicles to drive less. And, depending on their design, they may provide little or no incentive for improving new vehicle fuel economy.[7] Again, any climate-policy rationale for these taxes is removed with appropriate pricing of CO_2.

Revenue Use

Carbon taxes can provide a substantial new revenue source, which is especially valuable in times of fiscal consolidation. Revenues under an appropriately scaled carbon tax—about US$25 per tonne of CO_2 (see Chapters 3 and 4)—would amount to about 1 percent of GDP for many countries (and even more for fossil fuel–intensive economies like China, India, and eastern Europe). As a rough rule of thumb, up to 5 percent of this revenue might be required to administer the carbon tax. How should the rest of the revenue be used?

Earmarking of *all* tax revenues for environmental programs (e.g., subsidies for clean technologies, climate finance, research and development, or compensation for industry) is not generally desirable. The amount of revenue raised from a carbon tax has nothing to do with the socially desirable amount of spending on environmental programs. Instead, these programs need to be justified in their own right by additional market failures (see below); that is, they need to generate economic benefits comparable to those from alternative revenue uses.

The simplest way to use revenues to boost economic efficiency is to finance reductions in other taxes that distort the broader economy. For example, income, payroll, and general consumption taxes tend to (moderately) reduce

[6] For residential electricity taxes, it is important to recognize the distinction between excise taxes and value-added taxes. Typically, residential electricity consumption is included in the base of a general value-added (or sales) tax. This is entirely appropriate, as consumption of all household goods and services should be included in these taxes to avoid distorting the choice among different consumer products. Excise taxes, on the other hand, apply only to electricity and therefore raise the price of electricity *relative* to other final goods. This is generally undesirable when more effective instruments for reducing emissions (i.e., carbon taxes) are available.

[7] A recent trend has been to vary vehicle ownership taxes with engine size classes, or emission rates per mile. Although these tax systems provide some incentives for fuel economy improvements, they are not cost-effective. They tend to place too much of the burden of emissions reductions on shifting people into vehicles that are just below a higher tax bracket and too little on other opportunities, such as improving the fuel economy of vehicles that are a long way from the next, lower tax bracket. And they distort people's choices over different vehicles by causing a bunching of demand for vehicles classified just below the next (higher) tax bracket.

labor force participation and effort on the job, and shift production toward the informal sector. This is because they reduce the real returns to formal work effort (i.e., the amount of goods that can be purchased with earnings from a given amount of work hours). Similarly, taxes paid by firms on the return to capital investments and taxes paid by households on dividend income and capital gains tend to reduce capital accumulation below levels that would be economically efficient. Using carbon tax revenues to reduce these broader tax distortions produces economic benefits by improving incentives for (formal) work effort and capital accumulation. It also helps to "lock in" the carbon tax, as a future government that wished to abandon the tax would presumably have to impose other (politically difficult) tax increases elsewhere to make up for lost revenue.

Offsetting this revenue-recycling benefit, however, is the adverse effect on economy-wide employment and investment as overall economic activity contracts (slightly) with the impact of carbon taxes on energy prices and production costs. In fact, despite the potential for revenue recycling, the overall costs of carbon taxes are typically positive, though fairly small—about 0.03 percent of GDP for the average developed economy in 2020 for the scale of carbon price considered here.[8] Carbon taxes, therefore, still need to be justified on environmental grounds: Roughly speaking, an efficient tax system would charge CO_2 emissions for environmental damages and meet the government's remaining revenue requirements, mostly through broader fiscal instruments.

Carbon tax revenues might also be used for deficit reduction. Again, this use of revenues implies a significant economic benefit if it avoids the need for (near-term or more distant) increases in other distortionary taxes or helps to avoid economic crises. Revenues could also finance socially desirable public spending—in fact, the returns to public investments in education, infrastructure, health, and so on, in developing economies can be especially high if they suffer from capital scarcity (e.g., Collier and others, 2009). Revenues could also be used domestically for adaptation if the private sector would otherwise underinvest in these activities (e.g., defenses against higher sea levels would likely be inadequate without public support) or internationally for climate finance (see Chapter 7).

But the key point here is that revenues need to be used productively to keep down the overall costs of carbon taxes. If revenues are not used productively

[8] With revenues used efficiently, an (albeit rough) estimate of the annual economic costs of a carbon tax is given by one-half times the CO_2 tax times the emissions reduction. Suppose (from IMF, 2011) a US$25 per tonne CO_2 tax reduces OECD emissions by 10 percent from a base of 11 billion tonnes in 2020, then the approximate cost of the policy would be US$13.8 billion. Dividing by the projected OECD GDP of US$49.5 trillion (IMF, 2011) gives the above figure.

(e.g., if they are not used to lower tax rates to boost work effort or, worse, they are used for socially wasteful spending), this can greatly increase the overall cost of the policy to the economy. This is important to bear in mind if policymakers are considering using some of the revenues for compensation schemes (see below).

Nonetheless, there is a tension between environmental and fiscal objectives. The more effective a given carbon tax in reducing emissions, the more the tax base is eroded and the less revenue it will raise. On the other hand, even if the carbon tax is less successful in reducing emissions, the case for the tax may still be robust, as it provides a low-cost, relatively nondistorting way to raise public revenue.

Addressing Distributional Burdens and Industrial Competitiveness

The higher energy prices caused by carbon taxes are desirable to reduce emissions and promote clean technology investments. At the same time, however, they can have unpalatable implications for distributional incidence and industrial competitiveness, which can hold up the introduction of carbon taxes.

Given that—at least in middle- and high-income countries—spending on fuels and electricity tends to decline as a share of income as households become wealthier, poorer households tend to have the highest budget shares for energy, making them more vulnerable to higher energy prices.[9] In low-income countries, wealthier groups may be the most vulnerable, as the fraction of households owning vehicles or with access to the power grid may be much higher for them than for poor households (see Chapter 6), although wealthy groups are often politically powerful and successful in resisting energy taxes or securing exemptions from these taxes. In short, carbon taxes often run counter to distributional objectives, and in practice, their design may be subject to the constraint that they do not worsen income inequalities.

Higher energy prices may also harm the competitiveness of energy-intensive firms in trade-sensitive sectors where it is difficult to pass forward higher input costs into final product prices. Moreover, reduced production at home by these firms may cause increased production in other countries, thereby causing emissions leakage (i.e., emissions increases in other countries that offset some of the emissions reductions at home).

[9] One U.S. study found, for example, that the burden of a carbon tax on the bottom income decile is 3.7 percent of annual income, while it is a mere 0.8 percent of income for the top income decile (e.g., Hassett, Mathur, and Metcalf, 2009). The differential burden borne by low-income households may become less pronounced over time, however, as some of the burden of the carbon tax is shifted to owners of capital and of fossil fuels. And not all people with low income in a given year, who include, for example, college students, should be viewed as poor.

How might these problems be addressed? Artificially holding down energy prices (below levels warranted on environmental grounds) is not a good response, as most of the benefits leak away to other households and firms, rather than the target groups.

One way to alleviate concerns about incidence and competitiveness, and circumvent the pressure for border adjustments, is to scale back preexisting, environmentally ineffective energy taxes. In many OECD countries, the impacts of carbon taxes on electricity prices could, at least in part, be offset by lowering preexisting excise taxes on electricity use at the household level and in some cases at the industry level (IMF, 2011). In fact, with the pricing of both carbon and local air pollution in place, excise taxes on electricity become redundant (from an environmental perspective). Similarly, in many countries, the added burden of carbon pricing on motorists can be approximately offset by lowering vehicle ownership taxes (IMF, 2011).

Another approach is to alter the broader tax/benefit system, using some of the carbon tax revenue to approximately compensate target groups. In countries where low-income households pay either income taxes or payroll taxes, increasing the threshold income level below which no tax is paid provides a bigger rebate (relative to income) for these households compared with wealthier households (see the discussion of Australia's carbon pricing scheme in Chapter 8).[10] Moreover, cutting the average rate of income/payroll tax still has some favorable effects on work incentives. In countries where many households do not pay direct taxes, a possibility might be to use a portion of carbon tax revenue to finance a transfer to low-income households, and the rest could be used to finance a general reduction in consumption taxes. For vulnerable firms, transitory subsidies for production, or adoption of energy-saving technologies, might be provided to roughly offset the harmful effect of higher energy prices on competitiveness.

The danger of these compensation schemes, however, is that they can sacrifice some of the potential economic benefits from recycling carbon tax revenues. Transfer payments to low-income households, for example, do not improve work incentives. For the greatest economic efficiency, compensation would be kept to the minimum needed to offset the adverse distributional effects of carbon taxes for the target groups and, as much as possible, would take the form of tax cuts that alleviate broader distortions in the economy.

In principle, complementing a carbon tax with a well-designed and well-implemented system of border tax adjustments could be an effective way to deal with competitiveness and leakage issues. And border tax adjustments

[10] If people receive payments under an earned income tax credit scheme, heating supplements, and so on, the threshold could be the level of income above which no payments are received.

encourage other countries to participate in pricing regimes (as these countries are penalized for not joining).

Assuming border adjustments reflect genuine economic concerns rather than protectionist pressures from domestic interests, there are two practical implementation challenges. First, these adjustments could quickly become administratively complex if they are applied to many products and if rates are differentiated according to the carbon intensity of the exporting country. However, it makes sense to concentrate the adjustments on industries where competitiveness concerns are especially acute—mainly intermediate products like chemicals and plastics, primary metals (e.g., steel, aluminum), and petroleum refining. The second problem is that these adjustments might possibly (depending on how they are interpreted) run afoul of free trade agreements, in which case they may need to be designed somewhat differently (and less efficiently).

Yet another possibility is that if a carbon tax that limits harmful impacts on vulnerable firms and households is infeasible for the present, a series of tax-subsidy policies, known as "feebates," could be implemented. Chapter 1 explains how these policies work and how they might be applied to lower average CO_2 per kWh from power generation and improve the energy efficiency of vehicles, appliances, energy-using machines, and so on. Policymakers are free to choose "pivot points" for emissions intensity or energy consumption rates above/below which firms pay fees/receive rebates: A higher pivot point implies a smaller impact of the policy on energy prices (which may help with acceptability), although it also implies that less revenue will be raised (as a greater portion of firms receive rebates rather than pay fees).

From an environmental perspective, the drawback of feebates is that, unlike a carbon tax, they provide weaker incentives for conservation downstream from where the feebate is applied; for example, they do not encourage people to drive less and may provide little incentive to conserve on use of electricity-using products. Moreover, they cannot be implemented all the way upstream (e.g., on refineries), which raises administrative and compliance costs (see below). Nonetheless, feebates offer a reasonably effective and cost-effective way to exploit many (although not all) opportunities for reducing emissions that would be forthcoming under carbon taxes, while largely avoiding the need for compensating households or trade-sensitive sectors.

Finally, none of the above responses alleviates the burden of carbon taxes on upstream fuel suppliers, particularly domestic coal industries. Even though the tax on coal may be mostly passed forward in higher prices, the industry will still contract, leading to a loss of profits and employment and political pressure to not properly price coal. In fact (in the absence of widespread development and deployment of carbon capture and storage [CCS] technologies), a key purpose of a carbon tax is to promote a substantial shift away from coal. For this case, assistance might instead take the form of worker retraining and job relocation programs (this would likely absorb only a minor fraction of the carbon tax revenues).

Administrative and Compliance Considerations

The choice of collection points for the carbon tax should aim to maximize emissions coverage while minimizing administrative and compliance costs as well as the risks that people and firms will evade by paying the statutory taxes. In the latter regard, the tax should usually be imposed where the number of covered entities is smallest—most obviously, administrative and compliance costs are lower for upstream systems than downstream systems.[11]

Even under an upstream approach, there are a range of options. Regarding oil, in countries like the United States, there are far fewer petroleum refineries than oil wells, implying that the tax should be easier to collect at the refinery level. The (small) amount of oil used to make tar (which does not release emissions because it is not combusted) can easily be exempt from a refinery-level tax.

Natural gas is used mostly for heating residences and industry and for producing electrical power. Most natural gas comes from stand-alone gas wells, and a small amount is released from coal beds. Again, taking the United States as an example, it would make administrative sense to collect the carbon tax from approximately 500 of the largest operators—which would cover almost all of the reserves and production—rather than from the approximately 450,000 natural gas wells. A reasonable alternative to taxing the operators is to tax the processing plants plus the small amount of gas put into the pipeline system without processing.

As for coal, it is probably best to tax at the production level (mine mouth) rather than at the consumption level (electric utilities and industry) to limit collection points. In principle, the tax should vary moderately by coal type according to carbon content (anthracite, bituminous, subbituminous, and lignite emit 103.6, 93.5, 97.1, and 96.4 kg of CO_2 per million Btu, respectively), although administration may be easier without this differentiation.[12]

[11] For example, in the United States or the European Union, an upstream policy would apply to approximately 2,000 entities compared with about 12,000 entities in a downstream program.

[12] Cap-and-trade schemes often delegate the decision of what carbon contents to reckon for different grades of coal to a relevant agency, and the same agency could also do this for the carbon tax.

There should also be charges on carbon content at the seaports for the maritime contribution to refined oil products, natural gas, and coal.[13] Export taxes would be appropriate if domestic supplies are exported to regions that do not already price carbon.

In fact, for many developing economies, administering carbon taxes, which basically just requires monitoring fossil fuel supply, may be much easier than administering broader taxes, thereby strengthening the case for their inclusion as part of the broader fiscal system. For example, receipts from the personal income tax in developing economies tend to be low, reflecting the relatively large informal sector and tax evasion/avoidance opportunities for the wealthy.

Broader Coverage Issues

Energy-related CO_2 emissions account for about 80 percent of GHG emissions in developed economies (in CO_2 equivalents) and a somewhat smaller share in developing economies (where emissions from agriculture and deforestation are greater). Once CO_2 pricing is established, it makes sense to progressively expand the tax system to integrate other emissions sources, as institutional capability for reliable monitoring and verification is developed over time. The most urgent source of extensions (to the "low-hanging fruit" where potential emissions reductions are significant and reduction costs per tonne are relatively low) will vary by country: For some developing economies, sustaining forests may be the top priority.

Non-CO_2 GHGs, mostly methane, but also nitrous oxide, fluorinated gases, and sulfur hexafluoride, are relatively cheap to avert. Some of these sources would be fairly straightforward to include under a formal GHG tax, such as vented methane from underground coalmines and landfills.

Other sources could be incorporated through offset programs, where the onus is on the individual entity to demonstrate valid reductions (e.g., capture of methane from livestock waste in airtight tanks or covered lagoons). Under a tax regime, the primary effect of offset provisions is to reduce overall emissions (for a given emissions price), while under an emissions trading program, the primary effect is to lower allowance prices in the cap (without affecting total emissions). Offsets are effectively a subsidy to a polluting industry, however, and they need to be carefully designed to avoid the risk of encouraging more production from that industry.

[13] This is not a problem because the number of seaports for each of these fuels is limited and becoming increasingly more limited due to the greater size of oil-, gas-, and coal-transporting vessels, which necessitates deep-water facilities.

In a few cases, emissions can be difficult to integrate under the carbon tax. Examples include methane from surface mines (emissions are difficult to capture as they are released as the overburden is removed) and fluorinated gas emissions due to leakage from, or inappropriate disposal of, vehicle air conditioners.

Domestic carbon sequestration projects (reducing deforestation, reforesting abandoned cropland, harvested timberland, etc.) can also be integrated into domestic tax carbon regimes through emissions offset provisions.[14] But assessing the true carbon benefits of such projects can be quite challenging (see Chapter 5). Moreover, sequestered carbon in trees is not necessarily permanent if trees are later cut down, decay, or burn, which requires assignment of liability to either the offset buyer or seller for the lost carbon. And forest conservation in one country could lead to increased land clearance and emissions elsewhere, such as through upward pressure on global timber prices. Thus, while studies suggest that forest sequestration is often a low-cost option for reducing CO_2, policymakers should proceed cautiously with the integration of this sector to avoid undermining both the effectiveness and credibility of the tax regime.

Fossil fuel suppliers could be allowed to obtain tax credits by purchasing emission offset projects in developing economies, such as through the clean development mechanism (CDM).This is common in cap-and-trade regimes to date, but less so under carbon tax regimes. Again, it can be challenging to verify whether a project (e.g., a solar energy plant) would have gone ahead anyway without the offset (especially when the offset payment is small relative to plant construction costs). Although international offset programs are a potentially attractive way to channel funds for clean technologies to developing economies, again they should be integrated progressively under carbon tax regimes as the credibility of offset programs is established (it is not clear that the capacity of the CDM is large enough at present, however, to take on these extra duties).

Are Technology Policies Needed to Complement a Carbon Tax?

The ultimate objective is to switch from using conventional fossil fuels to phasing in more and more carbon-free fuels (solar, wind, nuclear, geothermal, coal with CCS, etc.), along with more efficient use of energy, and perhaps leaving a much greater part of coal, oil, gas, tar sands, and shale gas reserves unexploited. If things are left to the market without any price on carbon

[14] In this context, an offset is a reduction in emissions from a sequestration project that can be purchased by entities formally covered by a carbon tax in return for a corresponding reduction in their tax liability. The offset program might be limited to major landowners (e.g., the major paper and forest product companies) to limit administrative costs.

emissions, zero-carbon fuels will be phased in too late. Establishing a credible future path for carbon pricing is the single most important policy for encouraging the needed technology investments. As discussed in Chapters 3 and 4, the standard recommendation is that this carbon price should ramp up progressively over time, at around 2 to 5 percent a year in real terms.[15] However, even an appropriate time profile for a carbon tax may not be sufficient to engineer the change-over to low-carbon technologies if market impediments hinder their development.

There is some debate about whether carbon taxes should immediately be set at a very high level upfront to redirect technical change and rapidly reduce the emissions intensity of the energy system (e.g., Aghion, Veugelers, and Serre, 2009; van der Ploeg and Withagen, forthcoming).[16] In principle, even if policymakers wished to kick-start green innovation, it would be better to target technological opportunities with specific additional incentives rather than providing equal, across-the-board incentives for all emission reduction opportunities (regardless of the market impediments to individual technologies). Nonetheless, it can be challenging to design supplementary technology policies (in which case, higher taxes can have some role in promoting more innovation).

These challenges include (among others discussed below) the difficulty of picking "winners," the possibility that subsidies will be captured by lobbies for yesterday's technologies, and the possibility that very long-range benefits might be foregone if policymakers are overly focused on near-term innovation subsidies (at the expense of providing a credible long-term carbon price). Nonetheless, to the extent these challenges can be overcome, technology policies have an important role to play in complementing carbon taxes, as they can be targeted to where the sources of additional market failures (i.e., underinvestments in clean technologies) are most severe. To better understand these issues, we distinguish private research and development (R&D) from technology deployment (basic energy research funded by governments is also important, although it is difficult to make general policy recommendations in this case).

Private (Green) Research and Development

Even with a carbon tax in place, it is most likely that research conducted by private firms into clean technologies would be inadequate. Most importantly,

[15] Eventually, the efficient carbon tax path may flatten as the cost of extracting conventional fossil fuels rises over time as they become depleted.

[16] Some have argued that stiff taxes are also warranted because, on ethical grounds, climate change damages to future (unborn) generations should be discounted at rates below market rates, implying that the present value of future climate damages is much higher (see Chapter 4).

innovators cannot appropriate all of the spillover benefits to other firms that might copy a new technology or use information embodied in the technology to further their own research programs. Although this problem applies to private sector R&D in general (justifying broad-based policies to encourage all R&D), the problem may be more severe for CO_2-reducing technologies where, due to uncertainty over future government's commitment to climate policy, innovators are unsure about longer-term demand for clean climate technologies. "Network externalities"can represent a further impediment to the market implementing technologies like CCS, where construction of pipeline infrastructure to transport captured CO_2 to storage sites can benefit other firms. Although stimulating energy-related R&D may crowd out socially productive R&D elsewhere in the economy, as scientists and engineers are diverted from other sectors, full crowding out seems unlikely.

It is not entirely clear which type of technology instrument should be used. As already noted, it may be politically difficult to efficiently allocate large upfront subsidies for R&D across different technological opportunities. Strengthening patent protection (by increasing their duration or defining them more broadly) is another possibility, especially if the private sector knows more about the potential for diffusing new technologies than the government. Green technology/innovation prizes could also play a role when imitation around patents is still easy, although this requires that governments have some sense of the potential market for the technology. Another possibility is for the government to pay the original innovator a fee each time the new technology is adopted by another firm (with the fee corresponding to the value of estimated emissions reductions from the technology), although if the technology is improved later by other firms, it is not clear which firm should receive future adoption subsidies.

In short, while targeted incentives for private R&D into clean energy technologies are a potentially valuable complement to carbon taxes, their design needs to be carefully assessed, and the appropriate instrument may be different for different types of technologies. Higher initial carbon taxes are a further option if the need for redirecting technical change from CO_2-intensive to carbon-free modes of production is strong.

Technology Deployment

After a new technology has been brought to market, there is a further set of impediments that may prevent full (socially efficient) diffusion of the technology, although analysts continue to debate the seriousness of these impediments. For example, households may be unaware of the lifetime energy savings from more energy-efficient vehicles or appliances, and firms experimenting with a new type of technology may fail to capture the benefits from their "learning-by-doing" to other firms adopting the technology later on.

There is a potentially important role for informational campaigns here: If the problem is that consumers are unaware of potential energy savings, governments can provide information (through advertising, for example) to address that problem. There may also be a role for additional policy instruments to push the market penetration of specific new technologies, but again these instruments need to be carefully designed. One issue is that future net benefits of new technologies are uncertain—there is a downside risk that their costs may turn out to be higher than expected relative to alternative technologies, perhaps because of changes in fuel prices or prices of materials needed to manufacture the technology. Pricing instruments (like feebates or technology adoption subsidies) are better able to handle this uncertainty than regulatory approaches that force market penetration regardless of future costs—under pricing approaches, the technology will not be adopted if costs are excessive despite the policy incentive.

Another issue is that deployment policies should be transitory and phased out as the technology matures and becomes widely used (despite opposition from lobbies that may have been built up to keep them). Ideally, this phase-out would be announced up front and could be a function of time (e.g., the policy lasts 15 years, regardless of how much deployment occurs) or of market performance (e.g., the policy ratchets down whenever penetration goals are met) or some combination of the two.

International Issues

Top-Down Approaches

So far, countries have negotiated over country-level emissions targets and over side payments. The big CO_2 emitters in the past have been the developed economies, but in the next few decades, emerging economies (e.g., Brazil, China, India, Russia) are likely to be responsible for an increasing share of global emissions. A credible and effective coalition for mitigating climate change must include at least the main emerging and populous economies of China and India.

Negotiations over country-level emission targets are often contentious, not least because countries may have generous provisions for questionable emissions offsets that may effectively relax their target. Furthermore, updating emissions quotas over time is challenging, as baseline emissions of CO_2 in the absence of an effective climate policy grow at different rates across countries.

It might be a little less challenging to reach an international agreement over a common CO_2 price (and the annual rate of growth in that price) than over numerous country-level quotas. Prospects for agreement might be enhanced further if the tax took the form of a carbon tax floor. Such a floor is attractive in that it provides some protection for countries willing to set relatively high carbon taxes, and it reduces downside risks to clean technology innovation.[17]

A possible objection is that countries may undermine the floor through "fiscal cushioning" (use of broader energy tax/subsidy provisions to undermine the effectiveness of the formal CO_2 tax) or manipulation of other policies (e.g., avoiding significant regulation of local air pollution from coal-burning plants or charging far-below-market royalties for fossil fuel extraction on public lands). This is a potentially major problem, but it should not be overstated. These other provisions are typically very blunt at targeting emissions compared with a well-designed carbon tax and therefore have only limited impacts on offsetting the CO_2 emissions reductions from the tax. Nonetheless, a global carbon tax agreement would need to include provisions (e.g., monitoring by an international body) to address potential attempts at cushioning.

Another possible objection to carbon tax agreements is that countries forgo control over annual emissions targets. Offsetting this argument is that future climate change is driven by historical atmospheric GHG accumulations over many decades rather than emissions in any one given year. Nonetheless, a possible compromise (if policymakers wish to retain some direct control over emissions) might be to combine carbon tax floors with "carbon budgets." This would leave countries with flexibility over their annual emissions (subject to imposing the tax floor), but their cumulated emissions over, say, a 10-year period, could not exceed a maximum allowable amount (requiring increases in their carbon tax if they are not on track to stay within the carbon budget).

Bottom-Up Approaches

In the absence of a formal international agreement, individual countries might initiate their own pricing programs, which subsequently might be harmonized with those in other countries. It is sometimes argued that permit trading schemes are better for promoting a "bandwagon effect" to ultimately bring countries together in an international climate agreement, given mutual gains from trading permits at harmonized prices. However, carbon taxes might provide a similar bandwagon effect if they include border tax adjustments for imports from nations that have not already implemented pricing policies. When a new country joins the carbon tax agreement, it would then get to

[17] A useful precedent may be the minimum level of excise taxes and VAT rates agreed upon in the European Union.

keep the tax revenue on its exports—previously this revenue accrued to governments of other countries in the pricing agreement through border adjustments on those products.

International Aviation and Maritime Emissions

Chapter 7 discusses the strong environmental and fiscal case for, and potential implementation of, charges on CO_2 for international aviation and maritime emissions. International coordination here is especially important due to the mobility of the tax base; for example, in shipping, it is generally easy to refuel in ports that do not levy fuel charges. And there is currently much interest in international transportation fuel charges as a source of revenue for climate finance. If an international agreement over these charging schemes could be reached in the next few years, including acceptable compensation schemes for developing economies, this would set a valuable precedent for the (far more challenging) task of developing a comprehensive CO_2 pricing agreement.

Examples of Operational Carbon Taxes

Several countries or regions already have carbon taxes in place (Chapter 8 provides an in-depth discussion of experience with emissions pricing policies more generally).

Often (as previously recommended), these taxes have been implemented in a revenue neutral fashion; that is, other taxes were reduced when the carbon tax was introduced. For example, a large portion of the revenues from carbon pricing in Australia will finance a substantial increase in personal income tax thresholds.

And in some cases, the scale of the tax seems entirely reasonable based on discussions in Chapters 3 and 4. Australia is again a good model as the emissions price in 2012 will be equivalent to about US$25 per tonne of CO_2 (though the policy is an emissions trading program that will later allow more volatility in emissions prices). And in 2008, British Columbia, Canada, implemented a revenue-neutral carbon tax of US$10.4 per tonne of CO_2 that has since risen progressively to about US$30 per tonne.

Nonetheless, there are some significant differences in carbon tax rates across countries, suggesting potential for gains in trade (e.g., better harmonization of tax rates across countries or allowing entities in higher price regimes to purchase emissions reduction credits from countries with lower prices). This even occurs within Europe. For example, since 2005, Denmark has implemented a pricing scheme on fossil fuel emissions corresponding to US$114 (€80) per tonne of CO_2. Since 1991, Norway has charged a CO_2 tax on fossil fuels (on top of the excise taxes on fuels) amounting to US$21 per tonne of CO_2, while also in 1991, Sweden implemented its carbon energy tax amounting to US$126 per tonne of CO_2. China is planning a modest tax equivalent to US$3 per tonne of CO_2, rising to US$8 per tonne on heavy industry for 2012, initially starting with several pilot cities.

Moreover, for various reasons—particularly exemptions and tax preferences in response to concerns about equity and competitiveness and the use of multiple, overlapping tax instruments—there is often considerable disparity in emission prices across fuel types and fuel users, even within a country (see, e.g., discussions in Sumner, Bird, and Dobos, 2011, and Parry, Norregaard, and Heine, 2012). Whether there is a strong case for leveling tax rates depends, however, on whether fuel types/users that are subject to taxes that are markedly different from the average are responsible for a significant share of emissions.

Conclusion

Carbon taxes are especially timely. Their widespread implementation would jump-start the (long overdue) need to begin comprehensively controlling and scaling-back global GHGs, while providing across-the-board incentives for developing clean technologies ultimately needed to stabilize the global climate system. At the same time, they provide a valuable revenue source for cash-strapped governments. In principle, carbon taxes are pretty straightforward to design and administer. Ideally, they would be proportional to the carbon content of fuels and generally imposed upstream in the fossil fuel supply chain, such as a natural extension of existing tax systems for motor fuels.

One argument against carbon taxes is that the revenues might be squandered (or worse, used to fund socially unproductive spending). However, while the revenue use provisions in legislation accompanying the tax cannot be predicted in advance, it seems less likely that revenues would be wasted in today's fiscal climate, given that many governments are imposing painful spending cuts and tax increases to get budget deficits under control. Another argument against carbon taxes is that influential industries will seek exemptions or compensation for the burden of the tax, while low-income households may be unduly burdened from higher energy prices. However, there are some promising ways to deal with these types of challenges, from scaling back preexisting (redundant) taxes in the energy and transportation system to reductions in the broader tax system targeted at poor households to production subsidies (and possibly border adjustments) to compensate vulnerable firms. Yet another objection is that governments forgo direct control over their country's emissions, although even this concern may be partly addressed through complementing the policy with maximum-allowable carbon budgets over a period of years.

References and Suggested Readings

For a general discussion of pricing policies to address global climate change, see the following:

Aldy, J. E., A. J. Krupnick, R. G. Newell, I. W. H. Parry, and W. A. Pizer, 2010, "Designing Climate Mitigation Policy," *Journal of Economic Literature,* Vol. 48, pp. 903–934.

For details on existing emissions pricing programs, see the following:

Sumner, Jenny, Lori Bird, and Hilary Dobos, 2011, "Carbon Taxes: A Review of Experience and Policy Design Considerations," *Climate Policy,* Vol. 11, pp. 922–943.

Parry, Ian W.H., John Norregaard, and Dirk Heine, forthcoming, "Environmental Tax Reform: Principles from Theory and Practice to Date," *Annual Review of Resource Economics.*

For a good discussion of administrative issues for carbon taxes, see the following:

Metcalf, Gilbert E., and David Weisbach, 2009, "The Design of a Carbon Tax," *Harvard Environmental Law Review*, Vol. 33, pp. 499–556.

For a discussion of linkages between carbon taxes and the broader fiscal system, see the following:

Goulder, Lawrence H., ed., 2002, *Environmental Policymaking in Economies with Prior Tax Distortions* (Northampton, Massachusetts: Edward Elgar).

For a discussion on the burden of carbon taxes on different household income groups, see the following:

Hassett, K. A., A. Mathur, and G. Metcalf, 2009, "The Incidence of a U.S. Carbon Tax: A Lifetime and Regional Analysis," *Energy Journal,* Vol. 30, No. 2, pp. 155–175.

For a discussion of the possible use of stiff carbon taxes to kick-start green technological innovation, see the following:

Aghion, P., R. Veugelers, and C. Serre, 2009, "Cold Start for the Green Innovation Machine," Bruegel Policy Contribution 2009/12 (Brussels).

van der Ploeg, F., and C. Withagen, forthcoming, "Is There Really a Green Paradox?" *Journal of Environmental Economics and Management.*

For a discussion of the potential value of funding public spending (with carbon tax revenue) in developing economies, see the following:

Collier, P., R. van der Ploeg, M. Spence, and A. J. Venables, 2009, "Managing Resource Revenues in Developing Economies," *IMF Staff Papers,* Vol. 57, No. 1, pp. 84–118.

For some discussion of appropriate policies for addressing CO_2 and other adverse side effects of vehicles, see the following:

Parry, Ian W. H., Margaret Walls, and Winston Harrington, 2007, "Automobile Externalities and Policies," *Journal of Economic Literature,* Vol. 45, pp. 374–400.

For a comparison of fiscal instruments for reducing emissions and raising revenue, see:

International Monetary Fund, 2011. "Promising Domestic Fiscal Instruments for Climate Finance." Background Paper prepared by the International Monetary Fund for the Report to the G20 on "Mobilizing Sources of Climate Finance." Available at: www.imf.org/external/np/g20/pdf/110411b.pdf.

CHAPTER

3 Emissions Pricing to Stabilize Global Climate*

Valentina Bosetti
Fondazione Eni Enrico Mattei, Italy, and Centro Euro-Mediterraneo per i Cambiamenti Climatici

Carlo Carraro
University of Venice, Italy

Sergey Paltsev and John Reilly
Massachusetts Institute of Technology, United States

Key Messages for Policymakers

- Without significant emissions mitigation actions, projected "likely" global atmospheric temperature increases by the end of the century are approximately 2.5° C to 6.5° C above preindustrial levels.
- Although there is much uncertainty, a global carbon tax starting at roughly US$20 in 2020 and rising at 3 to 5 percent per year should be in line with stabilizing atmospheric greenhouse gas (GHG) concentrations at 650 parts per million (ppm) or keeping mean projected warming to about 3.6° C. A starting tax of roughly twice this level would be recommended if the goal is to keep atmospheric GHG concentrations to 550 ppm or mean projected warming below 3° C.
- However, keeping mean projected warming to 2° C (or stabilizing atmospheric GHG concentrations at current levels of about 450 ppm CO_2 equivalent), the goal identified in the Copenhagen Accord (COP 15) and reiterated in the Cancun Agreements (COP 16) is highly ambitious and may be infeasible. Achieving this

*This chapter is based on a policy note prepared for the IMF Workshop on Fiscal Policy and Climate Mitigation on September 16, 2011, in Washington, DC. We are grateful to Michael Keen, Ian Parry, and all the participants of the workshop for comments and suggestions. The usual disclaimer applies.

target would require the future development and wide-scale deployment of (still unproven) technologies that, on net, remove GHGs from the atmosphere. The Copenhagen pledges for 2020 still keep the 2° C target within reach—should these technologies be successfully developed—but highly aggressive actions would be needed immediately after that.

- Even the 550 ppm target would become technically out of reach if action by all countries is delayed beyond about 2030. And required near-term emissions prices (in developed economies) consistent with this target escalate rapidly with delayed action to control emissions in developing economies. Postponing mitigation actions, especially in emerging countries where large portions of energy capital are being installed for the first time, can be very costly. Extra costs associated with the delayed actions escalate rapidly with the stringency of the target, and some more stringent targets become infeasible if action is postponed.
- To reduce the cost while achieving an equitable sharing of them, decisions about where emissions reductions are taken and how they are paid for should be separated. Emission mitigation should take place where it is most efficient. Equity considerations can be addressed through agreed upon mechanisms that result in transfers from those better able to pay to those with less ability to bear these costs. Negotiating such a transfer scheme is likely one of the most difficult aspects of reaching an agreement.
- Innovation, both on energy efficiency and alternative energy sources, is needed. Carbon pricing (e.g., carbon taxes or a price established through a cap-and-trade system) would provide a signal to trigger both innovation and adoption of technologies needed for a low carbon economy.

In this chapter, we discuss projected greenhouse gas (GHG) emissions pricing paths that are potentially consistent with alternative targets for ultimately stabilizing the global climate system at the lowest economic cost and under alternative scenarios for country participation in pricing regimes. The pricing projections come from models that link simplified representations of the global climate system to models of the global economy, with varying degrees of detail on regional energy systems. There is considerable uncertainty surrounding future emissions prices, given that different models make very different assumptions about future emissions growth (in the absence of policy), the cost and availability of emissions-reducing technologies, and so on. Nonetheless, projections from the models still provide policymakers with some broad sense of the appropriate scale of (near-term and more distant) emissions prices that are consistent with alternative climate stabilization scenarios and how much these policies cost.

In the next section we discuss where we might be headed in the absence of mitigation policy, in terms of future GHG emissions trends, and what these imply for the growth of atmospheric GHG concentrations and, ultimately, for the amount of likely warming over this century. We also discuss the benefits of different stabilization targets for atmospheric GHG accumulations in terms of potentially avoiding warming. The chapter then addresses projected emissions pricing, as well as the costs of mitigation policies, to meet stabilization targets in the ideal (but unlikely) event of early and full global cooperation and with efficient pricing across all emissions sources and over time. This is followed by a discussion of the implications of delayed emissions reductions by all countries compared with just developing economies. We briefly evaluate recent emissions reduction pledges by country governments in light of the climate stabilization goals. The following section discusses the distributional burden of mitigation costs across countries and the potential complications for negotiation of long-term climate policy. In the final section, we offer some thoughts on pragmatic policy steps in the near term.

Emissions and Warming Trends

Emissions

There have been many efforts to project future emissions trends, and the range of projections over the twenty-first century has been wide. GDP and population expansion are major drivers of future emissions growth, although the role of the latter will gradually fade with the projected stabilization of the world population in the second half of the century. Some factors tend to dampen future emissions growth, such as potentially rising fossil fuel prices and improvements in energy efficiency (e.g., cars that can be driven longer distances per unit of fuel or buildings that require less energy to heat them). What differs most across forecasting models—and causes the uncertainty affecting emissions projections—are assumptions concerning future GDP growth; the availability of fossil resources; the pace and direction of technical change, in turn affecting the cost of low-carbon technologies and the energy intensity of the economy; and flexibility of fuel and technology substitution within the energy-economic system. Whether and when governments of high-emitting countries undertake meaningful GHG mitigation measures is an additional uncertainty on top of the various economic forces.

In the absence of (significant) mitigation action, energy-related carbon dioxide emissions (the primary GHG) are projected to increase substantially during the twenty-first century. Figure 3.1 shows the range of projections in a recent model comparison exercise organized by Stanford University's Energy Modeling Forum (EMF-22), which engaged 10 of the world's leading integrated assessment models.[1] On average, fossil fuel CO_2 emissions will grow from about 30 Gt CO_2 in 2000 to almost 100 Gt CO_2 by 2100.

[1] See Clarke and others (2009); four of the integrated assessment models participated with two alternative versions for a total of 14 models.

Figure 3.1. Energy-Related CO_2 Emissions Projections over the Twenty-First Century

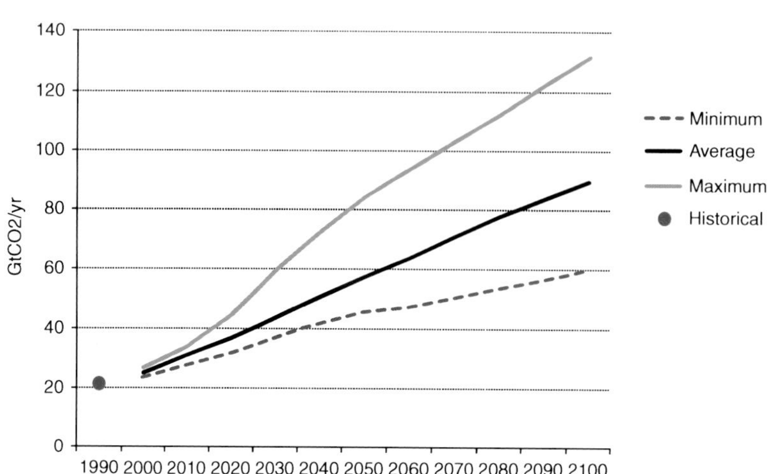

Source: Authors' calculations drawing from the EMF-22 dataset.

Note: The figure indicates a range of the median projections from each model used in the EMF-22 study.

The contribution of different regions to global CO_2 emissions is more stable across models. The Organization for Economic Cooperation and Development (OECD) countries will contribute 15 to 25 percent to total emissions in 2100 (compared with just under half of global emissions at present). Although the United States continues as one of the main emitters, its projected global emissions share will decrease from 25 percent to 10 percent over the century. Brazil, Russia, India, and China (BRIC) will contribute about 45 to 50 percent of total fossil CO_2 emissions by 2050, with at least 25 percent of the total emissions attributed to China alone from 2020 onwards. India accounts for a further 15 percent of global emissions by the mid-century. The rest of the developing world is projected to have an increasing role, moving from 17 to 25 percent of total emissions at present to 25 to 40 percent by 2100.

Anthropogenic CO_2 emissions are mostly energy related, with a (small) contribution from industrial processes (mostly cement production) and a (more substantial) contribution from land-use change, although energy-related emissions are projected to grow faster than these other sources of CO_2. Destruction of tropical forests and peat lands contributed 25 percent of global CO_2 emissions in 2000, mostly from a subset of tropical countries including Brazil, Indonesia, and some countries in central and western Africa.

While CO_2 is the major contributor to global warming, other GHGs also play a significant role. In particular, these include five other gases covered by the Kyoto Protocol: methane (CH_4), nitrous oxide (N_2O), and a group of so-called F-gases (HFCs, PFCs, and SF_6).[2] Currently, these non-CO_2 gases contribute about 25 percent of total annual GHG emissions in CO_2-e (i.e., CO_2 warming equivalents over their atmospheric life span), although again, these emissions are projected to grow slower than CO_2 emissions over the twenty-first century (IPCC, 2007).[3]

Implications for Future Atmospheric Concentrations and Temperatures

Once absorbed by the atmosphere, some GHGs are largely irreversible—CO_2 emissions, for example, reside in the atmosphere for about 100 years.[4] Without a significant emissions control policy, atmospheric GHG concentrations are projected to grow rapidly. The EMF-22 scenarios project atmospheric concentrations of 800 to 1,500 parts per million (ppm) CO_2-e by 2100 (counting concentrations of the gases identified for control in the Kyoto Protocol). For comparison, concentrations in 2010 were about 440 ppm.[5]

To date, temperatures are estimated to have risen by approximately 0.75° C relative to preindustrial levels, with most of the warming attributed to atmospheric GHG accumulations as opposed to other factors like urban heat absorption, volcanic activity, and changes in solar radiation (IPCC, 2007). However, the full impact of historical concentrations has yet to be felt due to inertia in the climate system (gradual heat diffusion processes in the oceans slow the adjustment of temperatures to higher GHG concentrations).

According to IPCC (2007), in the absence of a GHG mitigation policy, projected "likely" temperature increases by the end of the century are in the range of 2.4° C to 6.4° C above preindustrial levels ("likely" refers to a 66 percent chance or greater). A recent study at the Massachusetts Institute

[2] The major sources of F-gases are air conditioning, semiconductor production, electrical switchgear, and the production of aluminum and magnesium.

[3] Other substances will also affect future climate. These include chlorofluorocarbons (CFCs), whose emissions were largely phased out under provisions of the 1987 Montreal Protocol, but remain in the atmosphere as a powerful contribution to warming, and other short-lived warming substances like ozone and particulates. These substances add about another 30 ppm to atmospheric CO_2-e. On the other hand, some substances, particularly sulfates, have a cooling effect through deflecting incoming sunlight.

[4] Methane lifetime is about 12 years, nitrous oxide about 115 years, while F-gases lifetimes are thousands of years.

[5] It is important to distinguish between the concentrations of all GHGs and a subset of the Kyoto gases. In 2010, the CO_2 concentration was about 385 ppm and the Kyoto gases concentration was about 440 ppm CO_2-e, while for all GHGs, concentration was about 465 ppm CO_2-e. For more discussion on this issue, see Huang and others (2009).

of Technology (MIT) with updated climate and socioeconomic parameters projected even more warming—a 90 percent chance that temperatures will rise by 3.8° C to 7.0° C by 2100 with a mean projection of 5.2° C (Sokolov and others, 2009).

Yet another recent and especially comprehensive study by Prinn and others (2011) put together findings from intergovernmental panels (represented by the IPCC); national governments (including selected scenarios from the U.S. government Climate Change Science Program [US CCSP]); and industry (represented by Royal Dutch Shell). Prinn and others (2011) estimate global temperature increases of 4.5° C to 7.0° C from current levels by 2100 in the absence of climate policy. There are many risks associated with higher levels of temperature increase, some of which (particularly the risk of abrupt climate change) are poorly understood (see Chapter 4 and IPCC, 2007).

Avoiding Warming under Different Climate Stabilization Targets

Stabilization of GHG concentrations at levels often discussed in international negotiations would require very substantial emissions cuts. As indicated in Figure 3.2, some of the more stringent targets are already exceeded or will be exceeded in the not-so-distant future. In particular, the 450 CO_2-e target for

Figure 3.2. Relationship between Different CO_2-e Concentration Targets and Projected Concentrations in the Absence of Mitigation

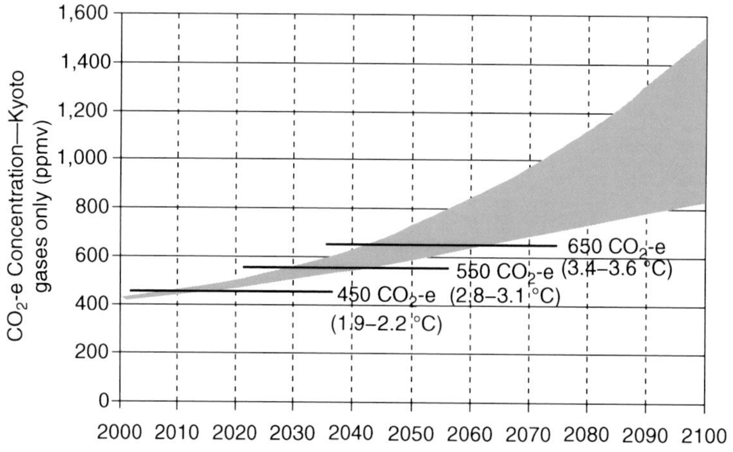

Source: EMF-22 (Clarke and others, 2009).

Note: Figures in parentheses indicate mean projected warming (above preindustrial levels) if concentrations are stabilized at particular levels assuming a value of climate sensitivity equal to three.

the Kyoto Protocol gases (consistent with keeping mean projected warming above preindustrial levels to approximately 2° C) is about to be passed.

Although the most stringent concentration targets might be beyond reach, even limited actions to reduce GHGs will appreciably reduce the probability of more extreme temperature increases. For example, according to results reported in Table 3.1, stabilizing GHG concentrations at 660 ppm rather than 790 ppm reduces the risk that warming in 2100 will exceed 4.75° C, going from 25 percent to less than 1 percent.[6]

But what scale of (near-term and more distant) emissions prices are needed to meet alternative stabilization targets and how much might these pricing policies cost? The answers depend, among other factors, on which countries participate in pricing regimes and the efficiency of the policies used to achieve emissions reductions. We turn to these issues in the next two sections, beginning first with the ideal global policy response with early and full participation in pricing regimes and then with more realistic scenarios with delayed action among all or a subset of countries.

Climate Stabilization with Global Participation of Countries

Here we consider a policy scenario with efficient (i.e., cost-minimizing) pricing of emissions across regions, different gases, and time, and full credibility of

Table 3.1. Cumulative Probability of Global Average Surface Warming from Preindustrial Levels to 2100

	ΔT > 2°C	ΔT > 2.75°C	ΔT > 4.75°C	ΔT > 6.75°C
No Policy at 1400	100%	100%	85%	25%
Stabilize at 900	100%	100%	25%	0.25%
Stabilize at 790	100%	97%	7%	<0.25%
Stabilize at 660	97%	80%	0.25%	<0.25%
Stabilize at 550	80%	25%	<0.25%	<0.25%

Source: Adapted from Webster and others (2009).

Note: Results are based on 400 simulations of the MIT's Integrated Global System Model under different assumptions about future emissions growth and parameters in the model. As the increase in global temperature from preindustrial levels to 2000 was about 0.75° C, probabilities for temperature increases relative to 2000 can be obtained by subtracting 0.75° C from the targets in the top row.

[6] The estimates in Table 3.1 should not be taken too literally, as they depend on assumptions about the probability distribution for warming at different long-term concentration levels, which are uncertain. The point is just that the risk of more extreme warming outcomes can be diminished sharply by stabilizing at lower GHG concentration levels.

future policies in triggering long-term investments. The ideal case is useful to understand the basic dynamics of the system and to have a benchmark for evaluating how far more realistic policy scenarios diverge from the ideal policy.

Achieving global economic efficiency (i.e., reaching a climate stabilization target at the lowest global economic cost) involves pricing emissions at the same rate across different countries. This can be achieved by imposing the same GHG price across the countries through a system of carbon taxes or by allowing a full trade in emissions permits among all countries and all sectors of the economy.

One caveat is that the focus here is on the total global costs of policies and not the cost that might be borne by individual countries. As discussed further below, there are numerous possibilities for sharing the burden of mitigation costs across different countries. Another caveat is that possibilities for reduced deforestation in tropical countries and reforestation of temperate regions are not captured in the model results discussed below, even though they could contribute significantly to mitigation efforts.[7]

Emissions Prices and Emissions Reductions

Projected emissions prices (in CO_2-e for all GHGs and in 2005 U.S. dollars) and CO_2 reductions for different concentration targets in the EMF-22 exercise are summarized in Table 3.2, where the ranges include cases that do and do not permit transitory overshooting of the long-term stabilization target.

The global emissions price in 2020 that would be in line with a 650-ppm CO_2-e target ranges between \$3 and \$20 per metric tonne of CO_2-e across the different models. Emissions prices in 2020 are \$4 to \$52 per tonne under the 550 ppm CO_2-e target (or \$10 to \$52 per tonne if overshooting is not permitted). For the 450 ppm CO_2-e target, only two models find a solution when no overshooting is allowed: with overshooting, half of the models are able to find a solution, although the 2020 emissions price is generally quite high at \$15 to \$263 per tonne.

The reason models are less capable of finding a feasible set of actions for more stringent targets is that concentrations are already very close to 450 ppm

[7] There may be significant and relatively low cost opportunities for reducing emissions through protecting and enhancing global forest carbon stocks. Reducing emissions from deforestation and forest degradation (REDD) could lower the total costs of climate stabilization policies by around 10 to 25 percent or, alternatively, enable additional reductions of about 20 ppm CO_2-e with no added costs compared to an energy sector–only policy (see Chapter 5 and Bosetti and others, 2011). However, implementation issues would need to be overcome (see Chapter 5): most of the rainforest countries have not yet developed the implementation capacity for monitoring and enforcing country-level projects, which might diminish the role of REDD in the next decade.

Table 3.2. Emissions Prices and Reductions under Climate Stabilization Scenarios: Full Global Participation

Atmospheric stabilization target, ppm CO_2-e	Emissions price in 2020 (2005 US$ per tonne of CO_2-e)[a]	Percent change in global CO_2 emissions in 2020 relative to 2000	Percent change in global CO_2 emissions in 2050 relative to 2000
450[a, b]	15-263	–67 to 31	–13 to –92
550[b]	4-52	–4 to 50	–67 to 52
650	3-20	30 to 57	–16 to 108

Source: Authors' elaboration of the EMF-22 dataset.
[a]Only a limited number of the models are able to solve for this case (even with overshooting) as it requires the development and wide-scale deployment of negative emission technologies.
[b]Includes cases both with and without transitory overshooting of the long-term stabilization target. In the 650-ppm case, there is no overshooting.

CO_2-e (Figure 3.2). Stabilizing at 450 ppm CO_2-e would require an immediate and almost complete decarbonization of the entire global economy, which is most likely technically infeasible.[8] Similarly, going back to the target after overshooting implies deployment on a massive scale of "negative emissions technologies" to remove CO_2 from the atmosphere, particularly biomass power generation coupled with carbon capture and storage (BECS). Not all models envision the future deployment of such technologies, which are highly speculative at present.[9]

For cost-effectiveness over time, the emissions price should rise at (approximately) the discount rate (or rate of interest) to equate the (present value) of incremental abatement costs at different points in time (in emissions trading systems the allowance price would increase over time at this rate if there is perfect substitutability of trading in emissions permits and other financial instruments). Different modeling groups assume different (real) discount rates, usually in the range of 3 to 5 percent, so the carbon price would also increase over time at this rate.

Looking at emission reductions (expressed as percentage changes with respect to 2000 emissions), which need to be in line with the different targets (see the second and third column in Table 3.2) for the near- and medium-term, there is not much difference in appropriate emission reductions for 550 and 650 ppm—but very large emission reductions are required, even in the short term, for the 450 ppm CO_2-e scenario.

[8] A small amount of GHGs can be emitted to offset the annual decay of GHGs in the atmosphere.
[9] Another negative emissions possibility is filters for direct removal of CO_2 out of the atmosphere, but these technologies (which are extremely costly and energy intensive at present) were not incorporated in the EMF-22 models.

Policy Costs

Ideally, mitigation costs would be measured by economic welfare losses (see Chapter 1 and Paltsev and others, 2009), although GDP losses are more commonly reported in climate policy models.

EMF-22 reports the net present value of GDP costs (discounted at 5 percent) in the range of $0 to $24 trillion (in 2005 U.S. dollars) for 650 ppm CO_2-e stabilization, in the range of $4 to $65 trillion (in 2005 U.S. dollars) for 550 ppm CO_2-e stabilization, and $12 to $125 trillion for 450 ppm CO_2-e stabilization, considering the full range of scenarios.

Figure 3.3 reports costs for each participating model for the three different stabilization levels and considering different levels of participation (full participation and delayed participation) and paths with and without overshooting of the target. Concentrating here on the full participation cases, costs, expressed as a percent of the present value of world GDP, are approximately 0.1 to 1.5 percent, 0.3 to 2.8 percent, and 2.7 percent for the three concentration targets, respectively.

US CCSP (Clarke and others, 2007) also reported the cost of climate policy as a percentage reduction in the global GDP, but rather than net present values, they reported the loss in different periods of time. The most stringent stabilization level in this study is roughly equal to 550 ppm CO_2-e (450 ppm when only CO_2 contributions are considered). The loss of the world GDP in comparison to a scenario with no climate policy is in the range of 1 to 4 percent in 2040 and 1 to 16 percent in 2100.

Emissions pricing will induce emissions reductions in the sectors where these reductions are cheapest. Models have different views about the timing of emissions reduction, but most of the projections agree that the power generation sector will be the first area where less-carbon-emitting (e.g., natural gas) or almost-zero-carbon-emitting technologies (e.g., nuclear, hydro, renewables) are introduced because of various economic substitutes that already exist in this sector.[10] Less-emitting technologies in transportation (e.g., gasoline/electric hybrid vehicles, more fuel-efficient conventional vehicles) and energy-saving technologies in buildings and industry are also promising, but they currently look more expensive. Substantial reductions in GHG emissions in agriculture and cement production are also costly, but to achieve climate stabilization, emissions from all sectors of the economy need to be reduced drastically.

[10] Jacoby, O'Sullivan, and Paltsev (2012) provide an assessment of the role of natural gas in a potential U.S. climate policy considering recent shale gas development.

Figure 3.3. Policy Costs for the EMF-22 Dataset by Model Run

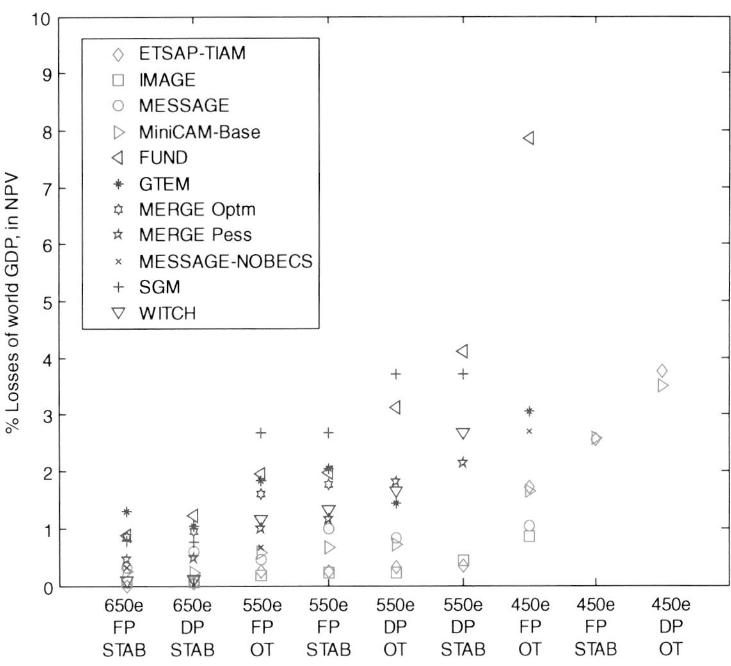

Source: Tavoni and Tol (2010).

Note: Green colors indicate models with biomass generation coupled with carbon capture and storage (BECS) and blue models without BECS. FP = full, immediate participation of developing economies, DP = delayed participation of developing economies, STAB = target not to exceed, and OS = target can be overshot.

For more stringent climate stabilization targets, these reductions obviously need to begin in the near future. Deferring the bulk of mitigation action to later periods can make sense if we are optimistic about the availability, cost, and speed of deployment of low-emission technologies. A further degree of freedom is represented by negative emissions technologies. However, relying on a technological future that might not evolve as expected comes at a risk of missing the target completely.

Delayed Action and Incomplete Participation

Here we discuss how delaying mitigation action and incomplete participation among countries in pricing agreements affects the feasibility of climate stabilization targets and the emissions prices and costs associated with those targets.

For a given stabilization target, delayed global action implies that once global GHG emissions have peaked, they must then be reduced at an even faster rate, which could require an abrupt and very costly replacement of capital. In fact, if the world continues according to business as usual until 2030, according to most models, stabilization at 550 ppm CO_2-e will no longer be possible (at least leaving aside highly optimistic scenarios for the development and deployment of negative emission technologies). This target might still be feasible if ambitious mitigation policies at the global scale are postponed until 2020, but this delay could substantially scale up global mitigation cost.

Rather than complete global inaction, more likely we will face asymmetry of actions across regions of the world. Significant mitigation actions are planned to take place in some developed economies within the next decade (e.g., the EU has pledged, by 2020, to reduce GHGs by 20 percent below 1990 levels). However, it is unlikely that emerging economies will make substantial emissions reductions in the coming decade.

Even asymmetric participation may rule out some of the more stringent targets, while scaling-up the global costs of those stabilization scenarios that remain feasible. Inaction in developing economies clashes with the fact that the bulk of emissions in the next decades will be coming from non-OECD countries.

If CO_2 emissions are not regulated in some major emitting countries, several inefficiencies arise. At a given point in time, low-cost mitigation opportunities in countries without a mitigation policy will go unexploited, while other countries must bear a greater burden of mitigation costs. A dynamic inefficiency also arises, as unregulated countries are those where most new investments will take place. Investing instead in fossil technologies, fast-growing countries may lock in these long-lived technologies (e.g., a new coal plant may be in use for 50 years), and later conversion to low-carbon technologies becomes more costly, or simply impossible, if early scrapping is deemed unfeasible. Finally, nonparticipating countries might react to lower fossil fuel prices (arising from decreased fuel demand in participating countries) and increase their emissions, thus partially offsetting the environmental benefit of early movers, though studies suggest this carbon leakage effect is not too large.[11] One solution, frequently discussed by economists, is the use of incentive systems (as for example an evolution of the Clean Development Mechanism) to induce reductions in developing economies while limiting leakage (e.g., Bosetti and Frankel, 2009).

[11] Most studies report carbon leakage from the Kyoto Protocol targets being in the range of 5 to 15 percent. See IPCC (2007), section 11.7.2.1 at: http://www.ipcc.ch/publications_and_data/ar4/wg3/en/ch11s11-7-2-1.html.

Figure 3.3 again reports results from EMF-22, which looked extensively into climate agreements with incomplete country participation. We now concentrate on the delayed participation cases for each stabilization level.

A key result is that even in the limited number of models that suggest that the 450 ppm CO_2-e stabilization scenario is feasible with early and full global cooperation over emissions mitigation (i.e., models with BECS technologies), the target becomes infeasible if only the OECD immediately undertakes mitigation action while BRICs and the rest of the world remain on their business-as-usual path until 2030 and 2050, respectively.

When participation of developing economies is delayed, half of the models cannot find a feasible set of investment actions that allow attaining the 550 ppm CO_2-e scenario. Still, with overshooting, the CO_2 emissions price faced by OECD countries in 2020 increases, on average, by a factor of three. On the other hand, delayed participation by developing economies does not make much difference to costs in the 650 CO_2-e stabilization scenario. In this case, there is much greater scope for additional mitigation by all countries in the second half of the century to offset the foregone reductions early in developing economies, while still keeping within the concentration target.

A further point from Figure 3.3 is the wide range of disagreement across models, depending on assumptions about flexibility of substitution across technologies and, once more, on the assumptions concerning the availability of BECS (green versus blue markers in Figure 3.3 distinguish models with and without BECS technologies).

The set of technologies that will be available and the speed at which they will be deployed can crucially affect not only the costs of any climate policy, but also the time we can wait while still staying on track with a climate stabilization target. The stricter the climate objective or the later the mitigation effort starts, the greater the need to develop technologies (such as BECS and CCS) that have potential implications that we have not yet fully understood. This obviously requires a careful and realistic estimation of the costs and potentials of these technologies, the research, development, and demonstration requirements to make them available with a reasonable level of certainty, and the potential barriers and possible adverse side effects (e.g., CO_2 leakage from storage sites) that might be linked to their deployment on a large scale.

How do projections we have discussed so far compare with the current state of climate negotiations? At the 2011 climate change meetings in Durban, South Africa (COP-17), for the first time, it was formally agreed that developing economies should be part of any future international emissions control regime (which is a step forward), although any control regime will not come into force before 2020 at the earliest. Prior to the meeting in Durban, countries agreed on submitting their emissions reduction "pledges" during the 2009 COP-15 in Copenhagen, Denmark, and the 2010 COP 16 in Cancun, Mexico, where most developed economies submitted their emissions reductions targets relative to emissions in 1990, 2000, or 2005.[12] Brazil, Indonesia, Mexico, the Republic of Korea, and South Africa proposed reductions relative to their business-as-usual emissions,[13] and China and India submitted carbon intensity reduction targets (i.e., CO_2 emissions per unit of GDP). Some of the pledges have conditions attached, such as the provision of finance and technology or ambitious actions from other countries, while some pledges were provided as ranges. This leads to some flexibility in their implementation and a range of potential outcomes.

Therefore, implications of these pledges for 2020 global emissions will depend on what pledges are implemented and what rules will be applied. Many scientific groups have estimated global emissions in 2020 based on the Copenhagen Accord pledges. The 2010 Emission Gap Report (UNEP, 2010) collected these estimates and showed that emissions in 2020 could be as low as 49 Gt CO_2-e (a range of 47 to 51 GtCO_2-e) when countries implement their conditional pledges in their more stringent form, or as high as 53 GtCO_2-e (a range of 52 to 57 GtCO_2-e) if pledges are implemented in their more lenient form.

Emission pathways consistent with a "likely" chance of meeting the 2° C limit generally peak before 2020, have emission levels in 2020 around 44 GtCO_2-e (a range of 39 to 44 GtCO_2-e), have extremely steep emission reductions afterward, and/or reach negative emissions in the longer term. Hence, the ranges implied by Copenhagen pledges do not necessarily rule out the 2° C target, as the two ranges are not severely distant from one another. However, as previously discussed, the larger the overshoot will be, the faster the decarbonization in the second half of the century will be needed, with all the implications that we have discussed in this chapter.

Who Bears the Costs of Abatement?

Distinguishing between who incurs mitigation costs and who actually implements mitigation activities is important. For example, mitigation might

[12] Typical targets for developed regions like Canada, the European Union, Japan, and the United States are in the range of 20 percent GHG reduction relative to 2000 levels.

[13] Targets expressed with respect to baseline emissions are particularly tricky as they can be interpreted in very different ways depending on the baseline projection adopted.

happen in developing economies but be financed by developed economies through an emissions offset program. Internationally allocating a given amount (typically determined by the stabilization target) of allowable emissions affects costs and who pays. This distributional issue would be extremely relevant both in the case of taxes and in that of permits. A number of possibilities for distributing the shares of emissions reduction among participating countries have been analyzed. Reductions might be based on equal percent reduction, GDP per capita, population, emissions intensity, historical responsibility, or many other alternative ways. As any of the schemes benefits (or imposes the cost on) countries unevenly in different aspects of socioeconomic indicators, there is no unique formula that would satisfy all participating countries.

There are two interacting equity concerns that would have to be dealt with in seeking the global emissions goal. First, incentives and compensation for developing economy participation will be required consistent with the principle of common but differentiated responsibilities. Second, since mitigation costs and compensation payments by developed economies will be substantial, they also will need to find an acceptable burden-sharing arrangement among themselves. Simple emissions reduction rules are incapable of dealing with the highly varying circumstances of different countries.

Successful climate negotiations will need to be grounded in a full understanding of the substantial amounts at stake. For example, for a 50 percent global GHG reduction by 2050 relative to 2000 (with full global participation in emissions pricing), Jacoby and others (2009) estimated that if developing economies (including China and India) are fully compensated for the costs of mitigation in the period up to 2050, then the average cost to developed economies is about 2 percent of the GDP in 2020 (relative to reference level), rising to 10 percent in 2050.[14] The implied financial transfers are huge—over US$400 billion per year in 2020 and rising to about US$3 trillion in 2050—with the United States' share amounting to US$200 billion in 2020 and over a trillion dollars in 2050.

It is surely extreme to assume that developing economies will demand complete compensation. Also, the amount of compensation is smaller if it only covers direct mitigation costs and not other losses, as might come through terms-of-trade effects.[15] Nonetheless, international financial transfers under more aggressive climate stabilization targets would remain of

[14] The required carbon prices in this exercise are rising from about US\$75/tCO_2 in 2020 to about US\$400/tCO_2 in 2050.

[15] In this case, the annual financial transfers to developing economies are lower by US$77 billion in 2020 and by US$108 billion in 2050 (Jacoby and others, 2009).

unprecedented scale and seem highly unrealistic (at least in the near term), given large budget deficits at present.[16] This further underscores the huge challenges to reaching a global agreement on rapid climate stabilization, challenges that only grow over time, when developing economies are expecting substantial compensation.

Yet another problem is that, besides being substantial, mitigation costs can also vary widely across countries. For example, mitigation costs are higher in energy-exporting countries, while energy importers have some terms-of-trade gains due to lower fossil fuel prices that allow them to reduce the cost of participating in emissions control regimes. Mitigation costs will also be greater in countries more dependent on carbon-intensive fuels and that employ inefficient mitigation instruments. Potential climate change damages also vary widely across countries but in a very different way. All these distributional impacts need careful study, as they complicate negotiation of long-term climate stabilization policy.

Conclusion

Advocates of rapid climate stabilization might be dismayed by some of the harsh technical, economic, and practical realities discussed in this chapter. Keeping mean projected warming above preindustrial levels to 2.0° C or stabilizing atmospheric GHGs at 450 ppm (about the current level) would require rapid widespread international adoption of emissions control policies and the development, and global deployment, of negative emission technologies later in the century to reverse atmospheric accumulations after a period of overshooting the long-term concentration target. Even the 550 ppm target (mean projected warming of 2.9° C) is extremely challenging, not least because required emissions prices escalate rapidly with further significant delay in controlling global GHGs, and the annual transfers to provide some compensation for developing economies are large and contentious to design. On the other hand, near-term emissions prices that are consistent with the 650 ppm target are more moderate, and delayed action on emissions reductions is less serious for this case, although obviously this target entails greater risks of dangerous warming.

The huge uncertainties—surrounding both the extent of climate change associated with a given atmospheric concentration target and our ability to develop technologies that would enable a rapid stabilization of the climate if the earth warms up rapidly—point to the importance of putting a policy architecture in place in the near term and delaying decisions about how

[16] Even one of the Copenhagen Accord goals of raising US$100 billion per year by 2020 for climate finance from "a wide variety of sources" seems extremely challenging at this point (see Chapter 7).

rapidly emissions should be scaled back in the distant future until some of the uncertainties have been resolved. Aiming for a CO_2 price somewhere in the ballpark of US$20 per tonne for 2020 applied across major emitting (developed and developing) countries seems reasonable and is roughly consistent with emissions prices suggested by the benefit-cost approach discussed in Chapter 4. Compensation issues for developing economies should also be more manageable at this level of pricing.

References and Suggested Readings

For details on the EMF-22 modeling exercise, see the following:

Clarke, L., J. Edmonds, V. Krey, R. Richels, S. Rose, and M. Tavoni, 2009, "International Climate Policy Architectures: Overview of the EMF 22 International Scenarios," *Energy Economics,* Vol. 31 (Supplement 2), pp. S64–S81.

For an overview of future emissions reduction pledges by different countries, see the following:

www.unep.org/climatepledges.org.

For a discussion concerning the potential role of bio-energy and carbon capture and storage (CCS) technologies on the costs of stringent policy, see the following:

Tavoni, M., and R. S. J. Tol, 2010, "Counting Only the Hits? The Risk of Underestimating the Costs of Stringent Climate Policy," *Climatic Change*, Vol. 100, pp. 769–778.

For a discussion about potential technological and economic obstacles for air capture technologies, see the following:

Ranjan, M., 2010, "Feasibility of Air Capture" (master's thesis; Cambridge, Massachusetts: Engineering Systems Division, Massachusetts Institute of Technology), available at http://sequestration.mit.edu/pdf/ManyaRanjan_Thesis_June2010.pdf.

Other publications:

Babiker, M., J. Reilly, and L. Viguier, 2004, "Is International Emissions Trading Always Beneficial?" *Energy Journal*, Vol. 25, No. 2, pp. 33–56.

Bosetti, V., and J. Frankel, 2009, "Global Climate Policy Architecture and Political Feasibility: Specific Formulas and Emission Targets to Attain 460 ppm CO_2 Concentrations," NBER Working Paper No. 15516, (Cambridge, Massachusetts: National Bureau of Economic Research).

Bosetti, V., R. Lubowski, A. Golub, and A. Markandya, 2011, "Linking Reduced Deforestation and a Global Carbon Market: Implications for Clean Energy Technology and Policy Flexibility," *Environment and Development Economics*, Vol. 16, No. 4, pp. 479–505.

Clarke, L., J. Edmonds, H. Jacoby, H. Pitcher, J. Reilly, and R. Richels, 2007, "Scenarios of the Greenhouse Gas Emission and Atmospheric Concentrations," Subreport 2.1A of Synthesis and Assessment Product 2.1 by the U.S. Climate Change Science Program and the Subcommittee on Global Change Research (Washington: Department of Energy, Office of Biological and Environmental Research).

Edenhofer, O., C. Carraro, J. C. Hourcade, K. Neuhoff, G. Luderer, C. Flachsland, M. Jakob, A. Popp, J. Steckel, J. Strohschein, N. Bauer, S. Brunner, M. Leimbach, H. Lotze-Campen, V. Bossetti, E. de Cian, M. Tavoni, O. Sassi, H. Waisman, R. Crassous-Doerfler, S. Monjon, S. Dröge, H. van Essen, P. del Río, and A. Türk, 2009, "RECIPE—The Economics of Decarbonization," synthesis report. Available at: www.pik-potsdam.de/recipe.

Huang, J., R. Wang, R. Prinn, and D. Cunnold, D., 2009, "A Semi-Empirical Representation of the Temporal Variation of Total Greenhouse Gas Levels Expressed as Equivalent Levels of Carbon Dioxide," Report 174 (Cambridge, Massachusetts: Massachusetts Institute of Technology, Joint Program on the Science and Policy of Global Change), available at http://globalchange.mit.edu/pubs/abstract.php?publication_id=1975.

IPCC, 2007, *Climate Change 2007: Mitigation of Climate Change.* Contribution of Working Group III to the Fourth Assessment Report of the Intergovernmental Panel on Climate Change, 2007. (New York: Cambridge University Press).

Jacoby, H., M. Babiker, S. Paltsev, and J. Reilly, 2009, "Sharing the Burden of GHG Reductions," in *Post-Kyoto International Climate Policy: Summary for Policymakers*, ed. by J. Aldy and R. Stavins (Cambridge, UK: Cambridge University Press), pp. 753–785.

Jacoby, H., F. O'Sullivan, and S. Paltsev, 2012, "The Influence of Shale Gas on U.S. Energy and Environmental Policy," *Economics of Energy and Environmental Policy*, Vol. 1, No. 1, pp. 37–51.

Morris, J., J. Reilly, and S. Paltsev, 2010, "Combining a Renewable Portfolio Standard with a Cap-and-Trade Policy: A General Equilibrium Analysis," Report 187 (Cambridge, Massachusetts: Massachusetts Institute of Technology, Joint Program on the Science and Policy of Global Change), available at: http://globalchange.mit.edu/pubs/abstract.php?publication_id=2069.

Paltsev, S., J. Reilly, H. Jacoby, and K. Tay, 2007, "How (and Why) Do Climate Policy Costs Differ Among Countries?" in *Human-Induced Climate Change: An Interdisciplinary Assessment*, ed. by M. Schlesinger, Haroon S. Kheshgi, Joel Smith, Francisco C. de la Chesnaye, John M. Reilly, Tom Wilson, and Charles Kolstad (Cambridge, UK: Cambridge University Press), pp. 282–293.

Paltsev, S., J. Reilly, H. Jacoby, and J. Morris, 2009, "The Cost of Climate Policy in the United States," Report 173 (Cambridge, Massachusetts: Massachusetts Institute of Technology, Joint Program on the Science and Policy of Global Change), available at http://globalchange.mit.edu/pubs/abstract.php?publication_id=1965.

Prinn R., S. Paltsev, A. Sokolov, M. Sarofim, J. Reilly, and H. Jacoby, 2011, "Scenarios with MIT Integrated Global System Model: Significant Global Warming Regardless of Different Approaches," *Climatic Change*, Vol. 104, No. 3-4, pp. 515–537.

Sokolov, A., P. Stone, C. Forest, R. Prinn, M. Sarofim, M. Webster, S. Paltsev, A. Schlosser, D. Kicklighter, S. Dutkiewicz, J. Reilly, C. Wang, B. Felzer, J. Melillo, and H. Jacoby, 2009, "Probabilistic Forecast for 21st Century Climate Based on Uncertainties in Emissions (Without Policy) and Climate Parameters," *Journal of Climate*, Vol. 22, No. 19, pp. 5175–5204.

United Nations Environment Programme, 2010, "The Emissions Gap Report: Are the Copenhagen Accord Pledges Sufficient to Limit Global Warming to 2° C or 1.5° C? A Preliminary Assessment" (Nairobi: United Nations Environment Programme).

Webster, M., A. Sokolov, J. Reilly, C. Forest, S. Paltsev, A. Schlosser, C. Wang, D. Kicklighter, M. Sarofim, J. Melillo, R. Prinn, and H. Jacoby, 2009, "Analysis of Climate Policy Targets Under Uncertainty" (Cambridge, Massachusetts: Massachusetts Institute of Technology, Joint Program on the Science and Policy of Global Change), available at http://globalchange.mit.edu/pubs/abstract.php?publication_id=1989.

CHAPTER

4 The Social Cost of Carbon: Valuing Carbon Reductions in Policy Analysis*

Charles Griffiths, Elizabeth Kopits, Alex Marten, Chris Moore, Steve Newbold, and Ann Wolverton
National Center for Environmental Economics, U.S. Environmental Protection Agency

Key Messages for Policymakers

- Without action to control rising greenhouse gases (GHGs), scientists predict that climate change will continue over time, bringing higher temperatures, sea level rise, and the potential for abrupt changes in earth system processes, with likely negative impacts on agricultural yields, ecosystems, human health, and more.
- These impacts are expected to vary widely over time and by geographic region, with developing countries likely to experience disproportionate damages due to limited adaptation opportunities and economic dependence on climate-sensitive sectors.
- The social cost of carbon (SCC) is the discounted monetary value of the future climate change damages due to one additional metric ton of carbon dioxide (CO_2) emissions.
- The U.S. government recently developed a set of SCC estimates for use in benefit-cost analyses of new regulations that influence CO_2 emissions—the central case estimate is $21 per tonne for emissions in the year 2010 (in 2007 U.S. dollars).
- Other countries could use these SCC estimates in benefit-cost analyses or to help set the initial level of a domestic carbon pricing policy, if those countries accept the policy judgments and methodological assumptions underlying the estimates.
- Other countries should consider developing their own SCC estimates when they wish to reflect fundamentally different assumptions, develop a long-term strategy to

* The views expressed in this chapter are those of the authors and do not necessarily represent those of the U.S. Environmental Protection Agency. We thank Joe Aldy, Terry Dinan, Robert Mendelsohn, Michael Keen, Ian Parry, and Tom Tietenberg for helpful comments and suggestions.

evaluate emission reductions, or evaluate the cost-effectiveness of policies designed to meet a given long-term target.

- The SCC estimates should be updated over time to reflect changes in emissions, atmospheric concentrations, economic conditions, and advancements in scientific knowledge.

Greenhouse gas (GHG) emissions from human activities—mainly from the burning of fossil fuels, deforestation, and agricultural activities—are accumulating in the atmosphere and altering the earth's climate and other natural systems.[1] Between 1850 and 2005, the atmospheric concentration of carbon dioxide (CO_2), the predominant anthropogenic GHG, increased from about 280 to 380 parts per million (ppm). Along with increased atmospheric accumulation of other GHGs, this has significantly contributed to an estimated increase in the global average annual surface temperature of approximately 0.8° C above preindustrial levels. Rising temperatures are also causing the level of the oceans to rise—on average by about 20 cm during the twentieth century. In the absence of serious policy action to abate GHG emissions, atmospheric CO_2 concentrations are projected to continue rising, with the potential to increase the global average annual surface temperature by an additional 1.1° C to 6.4° C and the average sea level by an additional 20 to 60 cm by the end of this century. Even if atmospheric GHG concentrations were to be stabilized immediately at 2000 levels, scientists estimate that the average surface temperature would likely increase by an additional 0.3° C to 0.9° C by the end of the century due to a delayed response inherent in the climate system (see Chapter 3 for further discussion of emissions and climate trends). Temperature increases and changes in precipitation will not be uniformly distributed across the globe—the specific magnitude, direction, and spatial pattern of these changes are highly uncertain and are an area of ongoing scientific research.

In addition, the scientific literature has paid increasing attention to the likelihood and nature of potential "climate catastrophes"; that is, high-impact,

[1] Fossil fuel combustion contributes about 26 Gt of CO_2 a year, and land-use changes contribute another 6 Gt. Agricultural activities are the primary source of other potent GHGs such as methane and nitrous oxide. Figures cited in this chapter are from the Intergovernmental Panel on Climate Change's Working Group I, "Summary for Policymakers" (IPCC, 2007).

low-probability earth system changes due to rising GHG concentrations. Possibilities include the collapse of the Greenland or West Antarctic ice sheets, a shutdown or change in the Atlantic Ocean circulation, substantially altered periodic weather patterns, large releases of additional GHGs from methane deposits, massive dieback of tropical or boreal forests, and cascading effects in marine food webs from ocean acidification. However, research quantifying the magnitude and timing of the physical—and especially the economic—impacts from many of these potential risks is still in its infancy.

Current and anticipated climate changes are expected to have a wide range of mostly negative impacts on economies and societies across the globe, including (but not limited to) the inundation of coastal areas, reduced agricultural yields, and increased frequency and severity of tropical storms, droughts, and other extreme weather events. Recent studies suggest that a 2.5° C to 3.0° C increase in the average annual surface temperature above preindustrial levels will lead to aggregate annual damages of between 0 and 2.5 percent of global gross domestic product (GDP). These aggregate figures mask considerable disparity in regional effects. One study estimated damages from a 2.5° C increase in global average temperature ranging from a positive 0.7 percent of GDP for the former Soviet Union to a negative 5 percent of GDP for South Asia and India (Nordhaus and Boyer, 2000). Likewise, a recent assessment found that a global sea level rise of 0.5 m to 2 m could displace 72 to 187 million people over the twenty-first century (assuming no adaptation), about 70 percent of which would be concentrated in east, southeast, and south asia (Nicholls and others, 2011). Developing countries will likely experience larger than average damages due to limited adaptation opportunities, as well as greater dependence of their economies on climate-sensitive sectors like agriculture, although nations' vulnerability to climate change may diminish with economic growth over the coming centuries.

With the aid of integrated assessment models (IAMs) that combine simplified representations of the climate system, the global economy, and their interactions, analysts can evaluate the economic implications of GHG mitigation policies. In this chapter, we describe recent efforts by the U.S. government to estimate the social cost of carbon (SCC) to value the damages from small changes in CO_2 emissions for use in benefit-cost analysis of policies that directly or indirectly reduce GHG emissions. We then discuss the potential applicability of these estimates to policy analysis in other countries and regions. We close by highlighting areas of future research and the need for frequent reassessment of the SCC in light of new information.

The Social Cost of Carbon

Defining the SCC

Comparing the benefits of policies that result in carbon reductions to their economic costs requires a monetized measure for the value of future climate change damages. The SCC is one such measure. It is the present value of future damages associated with an incremental increase (by convention, 1 metric tonne) in CO_2 emissions in a particular year expressed in consumption equivalent terms. In theory, it is intended to be a comprehensive measure including, for example, damages from changes in agricultural productivity, human health risks, property damages from increased flood frequencies, and the loss of ecosystem services.

The SCC can be calculated from a global perspective, which incorporates damages to all countries caused by CO_2 emissions, or from a domestic perspective, which incorporates only the damages experienced by a country's own residents. The U.S. government interagency working group chose a global perspective to evaluate the SCC, mainly because climate change is a global externality (CO_2 emissions rapidly become well mixed in the atmosphere and therefore contribute to damages around the world no matter their source). The use of a global SCC is consistent with the goal of achieving a globally economically efficient solution. If nations were instead to design their policies independently using domestic SCC estimates (thereby excluding benefits that accrue to nonresidents), abatement would be meaningfully less than the globally economically efficient level.

Integrated Assessment Models

Estimation of the economic impacts of CO_2 emissions can be broken into four main steps: (1) projections of future GHG emissions, (2) the effects of past and future emissions on the climate system, (3) the impact of changes in climate on the physical and biological environment, and (4) the translation of these environmental impacts into economic damages, discounted back to the present. IAMs couple physical science and economic models to capture important interactions between these components. The IAMs most often used to estimate the SCC represent highly aggregated, reduced-form approaches. While other more comprehensive models, such as computable general equilibrium models, may better represent the complex interactions among sectors of the economy and trade flows among countries, they typically lack the links between physical impacts due to climate change and economic damages necessary for estimating the SCC.

Most of the published SCC estimates are derived from one of three IAMs: William Nordhaus' Dynamic Integrated Climate Economy (DICE) model, Richard Tol's Climate Framework for Uncertainty, Negotiation, and Distribution (FUND) model, and Chris Hope's Policy Analysis for the Greenhouse Effect (PAGE) model. Each model takes a somewhat different approach to translate changes in climate variables, such as temperature and sea level, into economic damages. Some of these differences arise from the model developers' choices about the level of aggregation across regions, the climate variables, and which damage categories are explicitly included in each model (see Box 4.1).

Box 4.1. Damages in the PAGE, DICE, and FUND Integrated Assessment Models

As summarized in Table 4.1, the three integrated assessment models (IAMs) used for estimating the social cost of carbon (SCC) vary in their treatment of damages. For instance, the Dynamic Integrated Climate Economy (DICE) model aggregates damages across sectors and regions into a single global damage function, while the Policy Analysis for the Greenhouse Effect (PAGE) model divides damages into economic and noneconomic categories. In contrast, the Climate Framework for Uncertainty, Negotiation, and Distribution (FUND) model separately calculates damages for 11 different market and nonmarket categories. DICE and PAGE also include categories that account for the possibility of large "catastrophic" damages at higher temperatures, while FUND does not. In DICE, parameters are handled deterministically and represented by fixed point estimates; in PAGE and FUND, most parameters are represented by probability distributions. Also, unlike PAGE, DICE and FUND treat GDP as endogenous so damages in early years reduce economic output in later years.

The economic damage estimates vary considerably across models at both lower and higher global average temperature changes. This reflects differences in assumptions about the rate of technological change, the ability of human and natural systems to adapt to the effects of climate change, and the projected vulnerability of developing countries, among others. For instance, FUND projects that climate change is potentially beneficial for a 2.5° C increase due to effects on agriculture and forestry and decreased heating costs. PAGE assumes that impacts occur only above some "tolerable" temperature increase, defined as 2° C for developed countries and 0°C in developing countries. Beyond the tolerable level, developed countries are able to eliminate almost all economic impacts through various adaptation measures (e.g., altering crop varieties or planting dates, building sea walls), while developing countries can eventually eliminate 50 percent of economic impacts through adaptation. Adaptation to noneconomic impacts (e.g., biodiversity loss) is much more difficult for both regions. DICE does not include explicit representation of adaptation,

Box 4.1. (*continued*)

Table 4.1. Summary of Climate Change Impacts by Category

Global Damages at 2.5° C Above Preindustrial Levels (Percent of global output)					
DICE 2007		**PAGE 2002**		**FUND 3.5**	
Agriculture	0.13	Economic impacts	0.36	Agriculture and forestry	–0.90
Coastal	0.32	Noneconomic impacts	0.65	Coastal	0.02
Other market sectors	0.05			Hurricanes and other storms	0.01
Health	0.10			Health	0.07
Nonmarket amenities	–0.29			Water resources	0.16
Human settlements and ecosystems	0.17			Biodiversity Loss	0.13
				Cooling	0.90
				Heating	–0.51
Subtotal:	*0.48*	*Subtotal:*	*1.01*	*Subtotal:*	*–0.13*
Catastrophes	1.02	Catastrophes	0.43		
Total	***1.50***	***Total***	***1.44***	***Total***	***–0.13***

Sources: Based on Hope (2006), Nordhaus and Boyer (2000), and Tol (2009).

Note: For general illustrative purposes only based on default assumptions in DICE 2007 (output-weighted damages) and deterministic runs of FUND 3.5 and PAGE 2002 using the mode values for all parameters. Damage categories are defined by model authors. Refer to documentation for each of the three models for more detailed information on how these are defined.

though some forms of adaptation—especially in the agricultural sector—are included implicitly through the choice of studies used to calibrate the aggregate damage function.

Source: Interagency Working Group on Social Cost of Carbon (2010).

While IAMs offer useful guidance about the effects of climate change on human well-being, modeling the complex systems involved often requires assumptions that cannot easily be verified based on historical evidence. For this reason, IAM results should not be interpreted as precise predictions of far future outcomes. This is well understood and is frequently emphasized by the IAM developers themselves, and IAMs are regularly updated as modelers revisit key aspects of their framework (e.g., damage functions, assumptions about adaptation, the representation of natural systems to reflect the evolving scientific and economic research).

SCC Estimates Used in U.S. Regulatory Impact Analysis

In 2009–10, the U.S. government formed an interagency working group to develop a set of SCC estimates to be used by all executive branch agencies to value reductions in CO_2 emissions in regulatory analyses. The purpose of this interagency working group was to improve the accuracy and consistency with which agencies value reductions in CO_2 emissions. Prior to this effort, SCC estimates were used in some but not all regulatory analyses, and the values employed varied substantially among agencies.

The interagency working group used the DICE, PAGE, and FUND models to estimate SCC values. The climate system submodels and the functions that map climate change to economic damages were left unchanged, but a common set of assumptions for three key inputs was used across all the models—socioeconomic and emissions projections, equilibrium climate sensitivity, and discount rates.

The interagency working group selected five scenarios of GDP, population, and GHG emissions projections from the 2009 Stanford Energy Modeling Forum exercise (EMF-22).[2] These scenarios spanned a range of emissions projections (including at least one case where the rest of the world takes significant action to reduce emissions) and plausible outcomes for future population and GDP. The motivation for this approach was to ensure that the GDP, population, and emission trajectories were internally consistent for each scenario considered. Across the five scenarios, atmospheric CO_2 concentrations in 2100 ranged from about 450 to 890 ppm (or 550 to 1130 ppm in CO_2 equivalent when other GHGs such as methane were included), the average percentage change in global per capita GDP ranged from 1.5 to 2.0 percent per year, and the percentage change in global population was 0.4 to 0.5 percent per year. Future projections of global GDP were based on combining regional GDPs using market exchange rates.[3]

[2] The socioeconomic scenarios are available at Stanford's Energy Modeling Forum, a group of well-regarded energy modeling teams from Asia, Australia, Europe, and North America. See: http://emf.stanford.edu/research/emf22/.

[3] While the EMF-22 models use market exchange rates (MER) to calculate global GDP, it is also possible to use purchasing power parity. Purchasing power parity takes into account that some countries consume very different baskets of goods, including domestically produced nontradable goods. MERs tend to make low-income countries appear poorer than they actually are. Because many models assume convergence in per capita income over time, use of MER-adjusted GDP gives rise to projections of higher economic growth in low-income countries. There is an ongoing debate about how much this will affect estimated climate impacts. Critics of the use of MER argue that it leads to overstated economic growth and hence significant upward bias in projections of GHG emissions and unrealistically high future temperatures. Others argue that convergence of the emissions-intensity gap across countries at least partially offsets the overstated income gap so that differences in exchange rates have less of an effect on emissions. Nordhaus argues that the ideal approach is to use purchasing power parity but recognizes the practical data limitations of doing so over long time periods, although he also notes that exchange rate conversion issues are probably far less important than uncertainties about population or technological change.

In reduced-form IAMs, the speed and magnitude of temperature change for a given emissions projection are strongly influenced by the equilibrium climate sensitivity (ECS) parameter, which represents the long-term climate response to atmospheric conditions that are similar to those that would be associated with an atmospheric CO_2 concentration of 550 ppm, ignoring all other relevant gases.[4] To represent the uncertainty of the responsiveness of the climate system to changing atmospheric conditions, the interagency working group used a probability distribution to represent ECS in all three models. The probability distribution was calibrated to the Intergovernmental Panel on Climate Change consensus statement about this parameter by applying three constraints—a median equal to 3° C, two-thirds probability that the ECS lies between 2 and 4.5° C, and zero probability that it is less than 0° C or greater than 10° C.[5]

Discount Rates Selected

Since the damages from a tonne of CO_2 emissions occur over many decades, the discount rate—which reflects the trade-off between present and future consumption—plays a critical role in estimating the SCC. For policies with both intragenerational and intergenerational effects, U.S. federal agencies traditionally employ constant real discount rates of 3 and 7 percent per year. However, discounting over very long time horizons raises exceedingly difficult questions of science, economics, philosophy, and law. Approaches for determining the discount rate for climate change analysis have been categorized as either "descriptive" or "prescriptive."

The descriptive approach is based on observations of people's actual behaviors, such as saving versus consumption decisions over time and allocations of investment among more and less risky assets. Advocates of this approach argue that because expenditures to mitigate GHGs are a form of investment, discount rates used to evaluate benefits from these expenditures should be based on market rates of return.

The prescriptive approach to discounting specifies a social welfare function that formalizes the normative judgments that the decision maker wants to incorporate into the policy evaluation; that is, how interpersonal comparisons

[4] Specifically, the ECS represents the increase in the annual global-average surface temperature from a sustained doubling of atmospheric CO_2 relative to preindustrial levels.

[5] The truncation at 10° C is reasonable considering the very long time lags associated with such high climate sensitivity values (i.e., such high temperature outcomes could only occur well beyond the relevant timeframe for policy analysis using the discount rates employed by the interagency working group).

of well-being should be made and how the well-being of future generations should be weighed against that of the present generation. Proponents of the prescriptive approach argue that various market imperfections (e.g., the absence of markets for very long-lived loans) make the market interest rate an unreliable measure of the appropriate trade-off between the consumption of present and future generations; instead, the discount rate should be specified partly based on ethical judgments about intergenerational equity. Often the rates recommended by the prescriptive approach are lower than those based on the descriptive approach.

The interagency working group drew on both approaches but relied primarily on the descriptive approach to inform the choice of discount rate. With recognition of its limitations, the interagency working group felt that this approach was the most defensible and transparent given its consistency with the standard principles of benefit-cost analysis. Regardless of the theoretical approach used to derive the appropriate discount rate(s), it is important to note the inherent conceptual and practical difficulties of adequately capturing consumption trade-offs over many decades or even centuries. In light of disagreement in the literature on the appropriate market interest rate to use in this context and uncertainty about how interest rates may change over time, the interagency working group used three constant discount rates—2.5, 3, and 5 percent per year—to span a plausible range.

Calculating the SCC

Four basic steps are required to calculate the SCC in a particular year *t*. First, each model is used to project paths of temperature change and aggregate consumption associated with the baseline path of emissions, GDP, and population. Second, each model is re-run with an additional unit of CO_2 emissions in year *t* to determine the projection of temperature changes and aggregate consumption in all years beyond *t* along this perturbed path of emissions. Third, the marginal damages in each year are calculated as the difference between the aggregate consumption computed in steps 1 and 2. Finally, the resulting path of marginal damages is discounted and summed to calculate the present value of the marginal damages in year *t*.

The steps above were repeated in each model for multiple future years to 2050. Because the climate sensitivity parameter is modeled probabilistically and because PAGE and FUND incorporate uncertainty in other model parameters, the final output from each model run represents a distribution over the SCC in each year. The exercise produced 45 separate distributions of the SCC for a given year, based on the three models, three discount rates, and five socioeconomic scenarios considered. To produce a range of estimates that reflects this uncertainty but still emphasizes the central tendency, the distributions from each of the models and scenarios were equally weighted and combined to produce three separate probability distributions for the SCC in a given year, one for each assumed discount rate.

Four SCC estimates were selected from these three probability distributions to reflect the global damages caused by one tonne of CO_2 emissions: \$5, \$21, \$35, and \$65 for 2010 emission reductions (in 2007 U.S. dollars). The first three estimates are based on the average SCC across the three models and five socioeconomic and emissions scenarios for the 5, 3, and 2.5 percent discount rates, respectively. The fourth value is the ninety-fifth percentile of the SCC distribution at a 3 percent discount rate and was chosen to represent potential higher-than-expected impacts from anthropogenic GHG emissions. Figure 4.1 illustrates where these values fall within the wider distribution of SCC values generated by the three IAMs at three different discount rates. Notice that the distribution is skewed toward high values of the SCC. The range of the distribution increases with lower values of the discount rate.

The SCC estimates also grow over time because future emissions are expected to produce larger incremental damages as the economy grows and physical and economic systems become more stressed in response to greater climatic change. These rates are determined endogenously by the models

Figure 4.1. Distribution of 2010 Social Cost of Carbon Values at Each Discount Rate

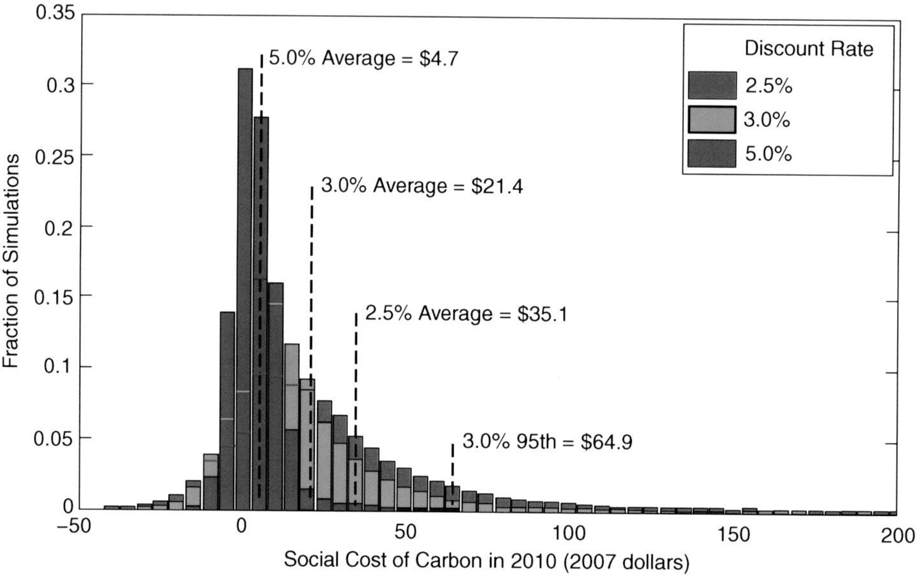

Source: Authors' calculations based on Interagency Working Group on Social Cost of Carbon (2010).

Table 4.2. Social Cost of Carbon, 2010–50 (In 2007 U.S. dollars per tonne of CO_2)

	Discount Rate			
Year of Emission Reduction	5 Percent Average	3 Percent Average	2.5 Percent Average	3 Percent 95th
2010	4.7	21.4	35.1	64.9
2015	5.7	23.8	38.4	72.8
2020	6.8	26.3	41.7	80.7
2030	9.7	32.8	50.0	100.0
2040	12.7	39.2	58.4	119.3
2050	15.7	44.9	65.0	136.2
Annualized percent change in SCC, 2010-50	3.1%	1.9%	1.6%	1.9%

Source: Modified from Interagency Working Group on Social Cost of Carbon (2010).

(see Table 4.2) and are dependent upon a number of assumptions including the socioeconomic and emissions scenario, model structure, parameter distributions, and the discount rate.

Having four estimates of the SCC raises the possibility that economic analyses of a single policy conducted with different values will generate different qualitative results. So, how should a policymaker interpret the results in such a case? In the United States, policymakers are asked to consider all four estimates of the SCC when conducting benefit-cost analysis, although the average values discounted at 3 percent are treated as central estimates. This is useful for purposes of informing decision makers of the robustness of a policy prescription to a range of values. It is also important to note that economic efficiency is only one of possibly many criteria U.S. policymakers consider when evaluating environmental policies. Furthermore, when there is substantial uncertainty surrounding the economic analysis, this may affect how much weight it is given relative to other criteria.

Using the Social Cost of Carbon for Policy Analysis

The appropriate role of the SCC in policy analysis and the applicability of the U.S. government's SCC estimates to analyses in other countries or regions depend on the nature of the policy in question, including the magnitude of the expected emission impact and timeframe of consideration. Because the SCC is the net present value of all future damages resulting from an additional tonne of CO_2 in year t, the estimates are conditional on forecasts of emissions and socioeconomic conditions from year t onward. Actions taken by the United States, or by other countries, to reduce emissions may change the forecasts of future damages. If these changes are large enough, then the future path of the SCC itself also would change. For this reason, the U.S. government's SCC estimates are most appropriate for analyzing policies that are expected to have a relatively small impact on global emissions and associated future climate conditions. To date, these values have been used to quantify the benefits of reducing CO_2 emissions from U.S. federal regulations such as energy efficiency standards for appliances and CO_2 tailpipe emission standards for light and medium heavy-duty vehicles. For long-term policies that have substantial impacts on global emissions, the appropriate SCC is one that accounts for the impact of the policy on forecasts of emissions and socioeconomic conditions.

Using the Social Cost of Carbon in Benefit-Cost Analysis

If policymakers in another country adopt the same set of policy judgments and methodological assumptions used by the U.S. government, then they can directly apply the SCC values described here to calculate monetized global benefits of CO_2 reductions resulting from domestic policies. The only adjustments needed involve translating the SCC estimates into the currency of the country where the analysis is to be conducted to allow for direct comparison to the domestic costs and other monetized benefits. In a developing country with a substantial portion of its economy made up of nontraded sectors, the market exchange rate may not provide an accurate assessment of what actions are worth taking, particularly when the SCC is being compared to domestic costs.[6] However, given that the SCC values estimated by the U.S. interagency working group are based on global GDP projections that use market exchange rates and the uncertainty involved in projecting purchasing power parity many years into the future, using market exchange rates is likely simpler and more transparent.

If policymakers in another country want to adopt other policy judgments or methodological assumptions, then the SCC estimates would need to be recalculated accordingly. Revised estimates could accommodate a variety of policy and ethical considerations, such as the use of a lower discount rate, used of a domestic SCC value, and equity weighting. We will now briefly discuss each consideration.

Discount Rates

While the interagency working group chose to use three different constant discount rates, other rates may also be consistent with what has been used

[6] This is not the case if a country uses the SCC values to approximate the level of a "global" carbon price. In this case, the ultimate objective of a social planner would be to maximize global benefits minus global costs. Any payments made across countries for emission reductions would occur at the market exchange rate.

in the literature. For instance, some decision makers may prefer to select a discount rate that reflects the prescriptive approach or a higher aversion to climate risks. Alternatively, they may want to use a nonconstant or declining rate to reflect greater uncertainty further into the future. If a different discount rate is desired to calculate the present value of avoided damages from one tonne of CO_2 emissions, then an entirely new set of SCC values must be generated. In general, use of a lower discount rate will result in a higher average SCC value. Incorporating uncertainty in the discount rate is also expected to increase the average SCC value relative to the equivalent constant discount rate (see Box 4.2 for a more detailed discussion of discount rate uncertainty).

Box 4.2. Treating Uncertainty in the Discount Rate over Long Time Horizons

While the U.S. interagency working group used a range of constant discount rates to generate SCC estimates, there is empirical and theoretical support for using a schedule of discount rates that declines over time. A number of studies in this area have found that uncertainty about future discount rates can have a large effect on net present value. A main result from these studies is that if there is a persistent element to the uncertainty in the discount rate, the effective discount rate declines over time. Consequently, lower discount rates tend to dominate the present value calculation over the very long term.

The proper way to model discount rate uncertainty remains an active area of research. One approach is to employ a model of how long-term interest rates change over time to forecast future discount rates. This type of model incorporates some of the basic features of how interest rates move over time, and its parameters are estimated based on historical observations of long-term rates. Subsequent work on this topic uses more general models of interest rate dynamics to allow for better forecasts that account for the present level of interest rates and the persistence of shocks. A simplified alternative to formally modeling uncertainty in the discount rate is to use a schedule of discount rates as has been done in both the United Kingdom and France. In this case, the analyst would apply a higher discount rate over the first 40 to 50 years of the policy and a graduated schedule of lower discount rates further out in time.

Source: Interagency Working Group on Social Cost of Carbon (2010).

Domestic SCC Values

Likewise, if policymakers wish to limit the analysis to a comparison of domestic benefits and costs, then the SCC values would need to be adjusted to exclude all damages experienced by residents of other nations. It is worth noting, however, that if each nation used an SCC estimate that included only its own domestic damages to evaluate regulations, then the result would be a much lower level of abatement in each country. Furthermore, the global level of abatement that would be realized under this scenario could be achieved at a lower cost if all countries used a common SCC value (or if international trading for emission rights was allowed) since marginal abatement costs would then be equalized across regulated sources.

Equity Weighting

Policymakers also may want to conduct a global social welfare analysis (using an explicit social welfare function that weighs both efficiency and distributional concerns) rather than, or in addition to, a benefit-cost analysis (which typically addresses economic efficiency alone). The SCC estimates developed by the U.S. interagency working group reflect an explicit decision to focus only on economic efficiency, which counts the willingness-to-pay of all individuals who will be affected by climate change equally, no matter their country of residence or income. It is possible to incorporate equity weights that adjust the measure of economic damages for differences in incomes among the affected individuals, but the SCC would need to be recalculated. In such an analysis, the monetized costs of the policy also would need to be adjusted using the same set of equity weights. In Box 4.3, we provide a brief perspective on the use of the SCC in other countries.

Box 4.3. An Illustration of Use of the SCC for Decision Making in Other Countries

The United States is not the only country that has developed social cost of carbon values. The United Kingdom first recommended the use of the SCC for informing national policies on GHG emissions in 2002 and commissioned a series of reports to examine the issue. The most widely known of these reports is the Stern Review. The SCC values estimated by Stern differ from the values used by the U.S. government in a number of ways. For example, Stern uses a lower discount rate and includes equity weighting. Relying on input provided by the Stern Review, the UK government officially set a value for the SCC in 2007 for the express purpose of determining the most appropriate limit on CO_2 emissions (referred to as a stabilization target when expressed in terms of the carbon concentration in the atmosphere). This value was set at about $50 per tonne of CO_2 equivalent in terms of 2007 U.S. dollars (i.e., £25.5 per tonne), rising at a rate of 2 percent annually. Germany followed the United Kingdom's

Box 4.3. (*continued*)

lead, relying on the same SCC value to evaluate its own domestic carbon policies, but suggesting sensitivity analysis using, in U.S. dollar terms, $15 and $215 per tonne.

In 2009, the United Kingdom moved away from using an SCC approach. It stated two reasons for this: (1) The SCC requires assumptions about what other countries will do to reduce GHGs and (2) there is a large degree of uncertainty around SCC estimates. In its place, the UK government now uses a shadow price of carbon for policy evaluation. It has two different sets of values, one to assess policies that reduce traded carbon emissions under the European Union's cap-and-trade policy and another to assess the cost-effectiveness of reducing GHGs in nontraded sectors of the economy.

Sources: DECC (2009); Umwelt Bundes Amt (2008).

SCC and Carbon Taxes

A country or group of countries may also be interested in using the SCC to help set the level of a carbon tax. However, it is important to note that the U.S. government's SCC values were calculated along a business-as-usual emissions path, not a socially optimal one. Therefore, countries should avoid locking in a long-term tax policy using the U.S. government's schedule of SCC estimates given in Table 4.2. Instead, the carbon tax in the near term should be set equal to the current best estimate of the global SCC and then adjusted over time to match an SCC that is reestimated as sources adopt new measures to reduce their emissions. Economic efficiency is achieved when the marginal abatement costs are equalized across all sources and are in turn equal to the current updated SCC.

In theory, a global carbon tax could be imposed at the outset without the need for adjusting the tax schedule over time, but only if the ultimate globally economically efficient level of emissions could be determined at the beginning. However, it should be noted that in the short run, the marginal benefit curve is relatively flat for CO_2 emission reductions because of the time lag inherent in the climate system and the relatively flat relationship between damages and climate change at relatively small temperature changes. This means that even for a nonmarginal tax policy, the SCC will not differ much from the marginal policy path in the near term. Figure 4.2 illustrates this point by showing the time path of the SCC along two separate emissions trajectories using the DICE 2010 model. The first represents a business-as-usual scenario. The second represents a large policy change, specifically a 50 percent reduction in CO_2 emissions each year relative to the business-as-usual path.

Figure 4.2. Social Cost of Carbon Values for Business-as-Usual and Nonmarginal Policy Emissions Paths

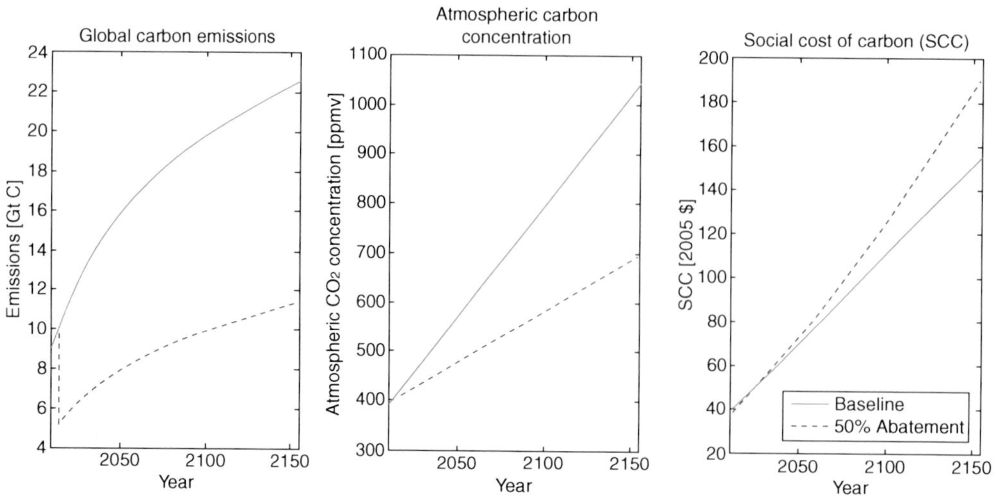

Source: Authors' calculations based on the DICE model.

Note: This example uses a constant consumption discount rate of 3 percent per year, but all remaining parameters are based on the default values in DICE 2010.

Notice that the SCC values for the two scenarios are very close to each other in the early years of the forecast: The SCC on the policy path remains within 5 percent of the SCC on the business-as-usual path for at least the first 45 years, even though the emissions on the policy path are 50 percent lower than those on the business-as-usual path starting immediately after the first time period. This is an extreme example, but it effectively demonstrates that it would be reasonable to use the SCC estimated along a business-as-usual forecast as a guide for setting a domestic carbon tax in the near term, such as in the next 10 to 20 years.[7]

Social Cost of Carbon and Cost-Effectiveness

The SCC is not the appropriate measure for evaluating projects when the overall policy objective is to meet a predetermined emissions (or concentration or temperature) target at the lowest possible cost. If the environmental target is specified ex ante, then a measure of the benefits

[7] Also note that in this example, the SCC increases when emissions are reduced. One reason this occurs is that in the DICE model, climate damages are represented as the loss of a fraction of gross economic output in each period, where the fraction of output lost depends only on the temperature anomaly in the period. Economic output net of climate damages is allocated to consumption and investment, so reduced damages in early periods can lead to increased economic output, and therefore increased absolute damages, in later periods.

of abatement is not needed. Instead, a measure of the marginal cost of abatement—sometimes called the "shadow price of carbon"—can be used to evaluate the cost-effectiveness of the policy (see Box 4.3). The two measures will be equal only when the emissions target is set at the economically efficient level.

Caveats and Reassessing the Social Cost of Carbon in the Future

The SCC estimates developed by the U.S. federal government for use in a benefit-cost analysis are subject to several important caveats. They are based on a number of assumptions and modeling simplifications that are not readily verifiable in the real world. For instance, there are many differences across the IAMs in how damages are modeled, as well as the treatment of technological change, adaptation, and catastrophic damages. Gaps in the literature make modifying these aspects of the models challenging, which highlights the need for additional research. Other key areas for future research include improvements in how predicted physical impacts translate into economic damages for a wide range of market and nonmarket damage categories, better incorporation of sectoral and regional interactions, the treatment of the discount rate in regulatory analyses where costs and benefits are widely separated in time (including how to address uncertainty), and methods for estimating the marginal damages from non-CO_2 GHG emissions.

In addition, the SCC estimates are based on a range of socioeconomic scenarios for how emissions will develop over time. As technologies develop, populations and economies grow, and countries formulate policies to reduce emissions, it is a virtual certainty that reality will diverge from what was assumed in the models to estimate the SCC. Thus, to keep pace with new research developments and to reflect diverging long-term trends, the SCC values should be reestimated on a regular basis.

Finally, it is worth emphasizing that attempts to estimate the SCC involve many unquantifiable uncertainties inherent to forecasting complex systems far into the future. This is due to the difficulty in accurately representing the many linkages between natural and economic systems as well as unforeseeable changes in population growth, technological progress, and regional economic growth. Therefore, the results from IAMs such as those discussed in this chapter should not be viewed as highly precise estimates of the SCC—or even precise estimates of the probability distribution over the SCC. Nevertheless, such models still provide invaluable information for policy analysis by quantifying what is currently known about potential climate damages and by providing a rigorous basis from which decision makers can assess the potential implications of the unavoidable modeling simplifications and omissions.

References and Suggested Reading

For background on climate trends and the science of global warming, see the following:

Intergovernmental Panel on Climate Change, 2007, "Summary for Policymakers," in *Climate Change 2007: The Physical Science Basis.* Contribution of Working Group I to the Fourth Assessment Report of the Intergovernmental Panel on Climate Change. (Cambridge, UK: Cambridge University Press).

On sea level rise specifically, see the following:

Nicholls, Robert J., Natasha Marinova, Jason A. Lowe, Sally Brown, Pier Vellinga, Diogo de Gusmão, Jochen Hinkel, and Richard S. J. Tol, 2011, "Sea-Level Rise and Its Possible Impacts Given a 'Beyond 4°C World' in the Twenty-First Century," *Philosophical Transactions of the Royal Society A*, Vol. 369, No. 1934, pp. 161–181. DOI: 10.1098/rsta.2010.0291.

For some discussion on the valuation of climate change damages and the social cost of carbon, see the following:

ICF International, 2011a, *Improving the Assessment and Valuation of Climate Change Impacts for Policy and Regulatory Analysis: U.S. EPA/DOE Workshop Summary Report. Part I: Modeling Climate Change Impacts and Associated Economic Damages*, http://yosemite.epa.gov/ee/epa/eerm.nsf/vwRepNumLookup/EE-0564?OpenDocument.

ICF International, 2011b, *Improving the Assessment and Valuation of Climate Change Impacts for Policy and Regulatory Analysis: U.S. EPA/DOE Workshop Summary Report. Part II: Research on Climate Change Impacts and Associated Economic Damages*, http://yosemite.epa.gov/ee/epa/eerm.nsf/vwRepNumLookup/EE-0566?OpenDocument.

Interagency Working Group on Social Cost of Carbon, 2010, *Social Cost of Carbon for Regulatory Impact Analysis under Executive Order 12866*, February, http://www.whitehouse.gov/sites/default/files/omb/inforeg/for-agencies/Social-Cost-of-Carbon-for-RIA.pdf.

National Research Council, 2009, *Hidden Costs of Energy: Unpriced Consequences of Energy Production and Use* (Washington: National Academies Press).

For more detail on the integrated assessment models used in the U.S. interagency working group report on social cost of carbon, see the following:

Hope, Chris, 2006, "The Marginal Impact of CO_2 from PAGE 2002: An Integrated Assessment Model Incorporating the IPCC's Five Reasons for Concern," *The Integrated Assessment Journal,* Vol. 6, No. 1, pp. 19–56.

Nordhaus, William, and Joseph Boyer, 2000, *Warming the World: Economic Models of Global Warming* (Cambridge, Massachusetts: MIT Press).

Tol, Richard, 2009, "An Analysis of Mitigation as a Response to Climate Change" (Copenhagen: Copenhagen Consensus on Climate).

For perspectives on discounting climate damages, see the following:
Portney, Paul, and John Weyant, eds., 1999, *Discounting and Intergenerational Equity* (Washington: Resources for the Future Press).

For other country perspectives on the SCC, see the following:
DECC, 2009, *Carbon Valuation in U.K. Policy Appraisal: A Revised Approach* (London: Department of Energy and Climate Change).

Umwelt Bundes Amt, 2008, *Economic Valuation of Environmental Damage: Methodological Convention for Estimates of Environmental Externalities* (Dessau-Rosslau, Germany: Federal Environment Agency).

CHAPTER

5 Forest Carbon Sequestration*

Robert Mendelsohn
Yale University, United States

Roger Sedjo
Resources for the Future, United States

Brent Sohngen
Ohio State University, United States

Key Messages for Policymakers

- An efficient forest carbon sequestration program could account for about a quarter of the desired global carbon dioxide (CO_2) mitigation over this century (with most of the remaining 75 percent from reducing carbon emissions from fossil fuels). An estimated 42 percent of this carbon storage could be achieved via reduced deforestation, 31 percent from forest management, and 27 percent from afforestation, with about 70 percent of overall carbon sequestration occurring in tropical regions.
- A serious deficiency in current sequestration programs is that each project is asked to prove additionality. However, it is not straightforward to identify which hectares are marginal and which would have stored carbon regardless. An administrative alternative is to establish a baseline level of carbon for forests in each country. Fees would then be charged for any reductions below the baseline and subsidies for carbon storage above the baseline. Setting the baseline equal to the existing level of carbon would lead to subsidies only for additional storage.
- Scaling-up small projects to promote forest carbon sequestration will be difficult given limited technical capacity and leakage. National programs are likely easier to administer.

* We thank Stephane Hallegatte, Alex Martin, Adele Morris, Sergey Paltsev, and Andrew Stocking for helpful comments and suggestions.

- National programs also allow national governments flexibility to address local institutions and property rights such as overlapping claims to timber, grazing, fuel wood, and nontimber forest products from the same forests.
- Measuring sequestered carbon can be problematic. Monitoring and enforcement is critical to maintain incentives for long-term storage. International agreements should encourage inexpensive monitoring technologies to keep these costs limited.
- The design of incentives is critical. For example, using forest coverage as a proxy for carbon storage provides no incentive to increase carbon per hectare. Similarly, lump sum payments for carbon give no incentive to protect established forests. Annual payments for the annual value of stored carbon encourage continued efforts to safeguard standing carbon.

Over a trillion tonnes of carbon dioxide (CO_2) are currently stored in biomass in the world's forests. Even without a carbon sequestration policy, forests appear to be sequestering an additional 4 billion tonnes of CO_2 per year. This net gain of 4 billion tonnes comes from a gross gain of 10 billion tonnes through forest planting and growth minus 6 billion tonnes of CO_2 lost from tropical deforestation each year. Some of this deforestation is harvesting for forest management, but a great deal of it is land conversion to agriculture. If one examines just the lost carbon from deforestation, forestry/land use causes 15 percent of man-made emissions. However, carbon cycle measurements confirm that forests are likely the sink for the 4 billion tonnes of "missing carbon." Whether forests can continue to be a sink depends upon the future effect of CO_2 fertilization and climate change on forest carbon stocks.

The key policy issue is not the baseline land-use emissions but rather what policy can do to increase carbon sequestration in forests. The Kyoto Protocol includes specific mechanisms to try to increase the stock of carbon in forests (Kyoto Protocol, Article 3.3). First, carbon storage can be increased by reducing deforestation. For example, the Forest Carbon Partnership Facility at the World Bank has developed a fund of about $400 million (World Bank, 2011) to reduce carbon emissions from deforestation. Second, carbon can be increased by planting trees in areas that are no longer forested (afforestation). Third, carbon storage can be enhanced by increasing forest intensity with plantations, fertilizer, or forest management.

How much additional carbon can be stored in forests depends on two things—what society is willing to spend to store more carbon and how quickly the carbon must be stored. The more rapidly carbon must be stored, the more expensive it is. A number of literature reviews have now shown that it may be possible to increase carbon in forests by about 4 billion tonnes CO_2

per year with a price per tonne of CO_2 of up to US$30 (in current dollars). This level of sequestration essentially doubles the net natural sequestration that is already occurring. Starting with an efficient price path for CO_2 from an integrated assessment model, an efficient universal program of forest carbon sequestration could account for 25 percent of all carbon mitigation (energy would be responsible for the bulk of the remaining mitigation). Many economic studies of carbon sequestration, however, have not addressed important administrative hurdles that a global program will have to face. Some of these, listed as follows, could be managed by carefully designing the sequestration program.

- "Leakage" can dramatically reduce the effectiveness of carbon sequestration if the program is not consistent across sites and is not universal.
- The process of storing carbon in forests is dynamic because it takes time for trees to grow and because the price of carbon changes over time. The sequestration instrument must be able to capture these dynamic properties.
- There are potential measurement and verification issues that need to be overcome to ensure that forests are being properly managed over time.
- Historically, carbon mitigation programs have tried not to pay for forest activities that might have been done anyway, and so they have been burdened by proving "additionality."

The following are also some problems that simply have not yet been addressed.

- Most analyses assume that forest carbon sequestration projects can be easily and quickly scaled-up from a few limited experiments to a globally comprehensive program with modest institutional costs.
- Forest ownership is often complex, especially in tropical countries. Many owners often have legitimate overlying claims on different forest amenities on the same piece of land.
- There are equity issues concerning who will be compensated by any carbon sequestration scheme.

We start this chapter by reviewing the potential of carbon sequestration in forests. The evidence suggests that forest sequestration is potentially an important source of mitigation. We then shift our focus to the administrative hurdles that must still be overcome to take advantage of carbon sequestration in forests.

Finally, we discuss the measurement and monitoring problems and the feasibility of scaling-up forest carbon sequestration globally in light of these complexities.

The Potential of Carbon Sequestration

Although some visionaries call for forests to be planted in deserts and other hostile locations, only a fraction of land is hospitable to forests. Growing forests in places without adequate soil and water would be prohibitively costly.

On lands that can support forests naturally, carbon sequestration can be achieved via three basic forestry activities—afforestation, forest management, and avoided deforestation. Afforestation involves converting former agricultural and abandoned crop lands back into forests. In areas where forests are most productive (i.e., moist tropical regions), they can sequester up to 11 tonnes of CO_2 per hectare per year in above-ground biomass and additional carbon below ground. Up to 2 billion hectares of forests have previously been deforested and converted to agriculture worldwide. All of this land could potentially be converted back to forests. Of course, this would leave us with little agricultural land. There is consequently a trade-off between forestland and agricultural land. The more land that is converted back to forestland, the higher the opportunity cost will be (from lost farmland). And carbon saved from afforestation takes a long time to be stored as it takes decades for trees to grow large enough to store substantial amounts of carbon.

Reducing emissions from deforestation is more promising. According to the Food and Agricultural Organization (FAO), an additional 6 million hectares of deforestation occurs each year globally, with most of the gross changes occurring in the tropics. Mature tropical forests contain a large stock of carbon per hectare (300 to 400 tonnes CO_2 per hectare). In the case of many tropical countries, a great deal of this stock is burned to prepare land for pasture or farming and therefore leads to a vast amount of immediate carbon emissions.

It is also possible to increase carbon in forests by changing forest management. According to the FAO, over 1 billion hectares of forests globally are currently production forests, but only 70 to 100 million hectares of forest are fast-growing plantations. Converting more forestland to plantations could quickly increase carbon sequestration. Additional potential management actions include postponing timber harvests, tree planting rather than natural regeneration, thinning to increase forest growth, fighting forest fires and other disturbances, and fertilizing.

If forest owners were paid US$30 per tonne of CO_2 permanently stored, they would be willing to sequester about 4 billion tonnes of additional CO_2

in forests each year. In an efficient program, approximately 42 percent of this carbon storage could be achieved via reduced deforestation, 31 percent via forest management, and 27 percent via afforestation. Afforestation accounts for relatively little of the additional carbon storage because it takes a long time for young forests to actually accumulate carbon and because the opportunity cost of forestland is high. In an efficient program, about 70 percent of carbon sequestration should occur in tropical regions (developing economies). Globally, 20 countries contain over 80 percent of the world's forest carbon. This group includes the five largest carbon-storing countries (Brazil, Canada, the Democratic Republic of Congo, Russia, and the United States) as well as Indonesia, Malaysia, and other countries in South America and Africa that are responsible for most of the global deforestation.

If forest owners were paid significantly more than US$30 per tonne of CO_2, they would be willing to store even more carbon. It is also true that forests could store more carbon if given more time. With more time, programs such as afforestation become increasingly effective. By 2100, for example, approximately 367 billion tonnes of CO_2 could be stored in forests cumulatively with a final price of US$50 per tonne CO_2, providing about 25 percent of the cumulative abatement over this period. With a price of US$110 per tonne CO_2, over 1.4 trillion tonnes could be stored by 2100 (Sathaye and others, 2006).

Institutional Hurdles

Scaling-Up

There are a host of small programs and case studies that have tried to reduce deforestation and increase afforestation in order to capture carbon in forests. Can these small projects easily be scaled-up to a global program in a decade? Past experience suggests that it is often very difficult to scale-up small experiments to even a national level, much less a global level. The experts and volunteer (nongovernmental) organizations that support all of these small-scale efforts are not sufficient to manage a global program. The program would have to expand to between 1,000 and 10,000 times its current size. The existing capacity could not manage such a vast increase. Many more people would have to be trained in forestry. It would take time and resources to increase the scale of current efforts.

Of course, scaling-up may be easier in some countries or regions than others. For example, the U.S. Conservation Reserve Program (CRP), which sets aside farmland in the United States for environmental protection, scaled up from 0 to over 12 million hectares in 5 years (U.S. Department of Agriculture, Farm Services Agency). The current administrative cost of the CRP is about $7 per hectare of land enrolled. This includes costs of administering the contracts and verifying that the practices are still in place in the 10–15 years of the contract length.

Although the CRP shows that it is possible in some circumstances to scale-up environmental protection programs relatively quickly, there were many complaints early on that the program did not pay full attention to environmental (primarily conservation) benefits. Many lands enrolled in the initial stages were low-value croplands in regions far from human population where the environmental benefits were less valuable. In addition, this program was conducted in the United States where land ownership is usually fairly easy to prove. An enlarged global forestry carbon program would require substantial attention be paid to the benefits of the program, and it would probably require clear titles to be obtained in regions where ownership may be disputed. These factors could make the program difficult to implement in many regions.

Dynamics

There are two reasons that a carbon sequestration program should be dynamic. First, the marginal benefit of carbon storage (social cost of carbon [SCC]) is the damage avoided by permanently sequestering a tonne of carbon. This marginal value increases over time as greenhouse gas concentrations rise. Consequently, the marginal cost of carbon sequestration programs should also rise over time as the marginal benefit rises. This makes the carbon sequestration program inherently dynamic, with more carbon being stored over time. Second, trees grow according to a sigmoid growth function (growth increases with age up to a maximum and then it decreases with age). Trees do not grow at a constant rate. Afforestation and forest management programs generate different amounts of carbon storage over time.

One way that carbon sequestration programs can be accurately tied to what each forest can provide is to rely on rental payments for annual carbon storage (rather than one-time payments for permanent storage; see, for example, Marland, Fruit, and Sedjo, 2001; and Sedjo and Marland, 2003). Using annual payments also provides a continued incentive for the forest owner to protect the forest. This is lost once an up-front lump-sum payment is made. Rental payments should equal the SCC (the present value of the stream of marginal damage caused by a tonne of carbon) times the interest rate. For example, with a real interest rate of 4 percent, the rental payment for an SCC of US$30 per tonne of CO_2 is equal to US$1.20 per tonne per year ($30 times 0.04).

Measurement

Measurements of forestlands and timber volumes have been under development for decades. For example, the U.S. Forest Service samples sites across the 700 million acres of U.S. forest every 5 years (although the exact sampling regimen varies by state). These ground measurements are then supplemented with aerial and remote sensing information. The FAO estimates global forest areas by country. Unfortunately, the quality of the data varies greatly across countries, so there is considerable uncertainty around their estimates. The total amount of land in forests is somewhat uncertain because there is a complicated edge between forested savannah and fully grown forests. However, the biggest uncertainty concerns the stocking per hectare of forests (the amount of carbon per hectare). This can vary by land productivity, by species, and by land management. For example, the annual sequestration rate for a typical New England forest is 0.5 tonnes per hectare per year, a southern pine plantation is 1 tonne, and a moist tropical forest could be as high as 11 tonnes.

It is somewhat easier to verify whether an acre of intact forest has been clear-cut. Satellite pictures over time can reveal dramatic changes in land cover such as a clear-cut. However, what is more difficult to verify is the biomass per hectare of forested land. The actual biomass is important because selective harvesting can reduce biomass without causing visible clear-cuts. Further, intensive forest management can increase biomass, but again, this is not visible to a satellite. Verification of the biomass per hectare in forests may require ground-truthing, which is very expensive. Current estimates of the monitoring costs of the U.S. system are $72 million per year, or $0.24 per hectare. The annual change in carbon in above-ground stocks is about 635 million tonnes of CO_2 per year, so the cost of measuring a change in carbon in forests in the United States is about $0.11 per tonne CO_2. This is relatively high compared to the annual value of a tonne of carbon storage, which is less than $1.20 per tonne.

The cost of monitoring is even higher with small isolated projects. Specific projects for smaller areas of 1,000 to 600,000 hectares could cost US$1 to US$2 per tonne CO_2 (see Antinori and Sathaye, 2007; and Antle and others, 2003). Measuring and monitoring regimens could be done every 5 years to keep these costs down. Measuring just above-ground carbon (usually about three-fourths of the total carbon) could also keep costs down. Some new promising technologies, such as Light Detection and Ranging (LIDAR), that rely on low-level aerial photography can estimate wood volumes much more cheaply than ground-truthing. However, the carbon content depends upon weight, not volume, and hence some activities in addition to LIDAR are required.

Additionality

The total cost of the carbon sequestration program depends not only on the price of carbon (rental rate per year), but also upon what carbon must be purchased. The simplest program to administer is to pay every forest owner the rental rate on every tonne of carbon stored. For example, if the rental rate is US$0.60 per tonne per year and there are 1 trillion tonnes of carbon stored in forests, that would involve a payment of US$60 billion per year every year. However, many architects of carbon policy wish to pay just for the additional carbon stored (not the baseline that would have been stored anyway). If a program stored 4 billion additional tonnes, that would require an annual payment of just US$0.024 billion. Only the additional tonnes would be paid for. Of course, this raises the intriguing question of what tonne is additional versus baseline. In practice, this is very difficult to determine, and past case study projects have been handicapped by proving additionality. It is very difficult for a project to prove what would have happened anyway and what will now happen with a carbon sequestration program. Would there, in fact, be a change in behavior because of the program or is there an incentive for every forest owner to simply claim it? It is very hard to identify the actions that are on the margin.

Other ways to avoid this problem with additionality involve switching the property rights. The rental methods described above assume that landowners or land managers have the right to sell carbon credits onto markets. The current policy discussion embraces this property right. However, society could instead decide to treat forests as a potential emission source and tax carbon emissions. A carbon tax at the time of timber harvest combined with a subsidy for annual growth would have the same overall economic costs as the carbon rental scheme described above, but it would not require society to determine additionality with each carbon project. A carbon tax and subsidy scheme would change the distribution of carbon payments, but it would not require spending resources to prove additionality. Of course, taxing forest owners for releases of carbon from their forests suddenly makes forests a liability. If not carefully handled, this could create perverse incentives for reducing forests even further prior to program implementation.

Leakage

Economic analyses of land use suggest a carbon storage program must be universal to be effective. Of particular concern is the global trade-off between forestland and farmland. If some lands are in the carbon storage program and some are not, the scarcity that the program lands create for farmland encourages non-program lands to convert forests to farming. This phenomenon is called leakage. It can dramatically reduce the effectiveness of the carbon storage program. For example, a reduction of timber harvests in one region may simply result in an increase in the market price and increased harvests elsewhere, either within the country but also perhaps beyond its boundaries. Also, suppose one set of countries joins the program and sets aside an additional 50 million hectares of land for carbon storage by converting farmland to forestland. This would dramatically increase the scarcity of farmland and create a huge incentive for the countries not in the program to convert their forestland to farmland. Depending on how substitutable the land might be, the nonparticipating countries could actually convert 50 million hectares of forestland to farmland in response to the incentives created by the program, making the carbon storage program completely ineffectual. Although this is a worst case scenario, the problem of leakage is not trivial.

One solution to leakage is a universal program. If all land everywhere faces the same incentive to store carbon, there would be no leakage. The carbon storage program does not technically have to be identical in every country. Some countries might use regulations or taxes, whereas other countries might be inclined to use subsidies and tax breaks. However, all of the programs must use the same effective marginal incentive to store carbon; otherwise, the leakage problem will reduce the effectiveness of the global effort. Of course, what is important is that most of the potential forestlands in the world face the same incentive. Therefore, what is really important is to have agreement across the countries with most of the world's forest. If the agreement could cover the 20 countries with the most forest in the world, about 80 percent of forests would be covered.

Some researchers have proposed discount factors to correct for potential leakage. Discount factors work by requiring suppliers of carbon credits to provide additional carbon for each credit claimed. For instance, if the discount factor is 2, a country would have to provide 2 units of carbon credits for each 1 unit that receives a payment. Discount factors penalize countries that engage in carbon sequestration by giving them rights to only a certain percentage of the carbon they could store, thus reducing the value of their carbon stock. Discounting for leakage raises costs arbitrarily, gives incentives for countries to remain out of the program, and creates other inefficiencies. When designing a carbon system, it is preferable to include elements that provide incentives for new countries to enter into the system and not for them to stay out of the system.

Permanence

The question of permanence arises because forests store carbon only temporarily, while the tonnes of carbon released into the atmosphere by energy processes are "permanently" added to the atmosphere. Forests planted expressly for carbon sequestration, for instance, will sequester and hold carbon only so long as they remain standing. There is some probability that forests will be affected by fires, pests, windstorms, human-directed harvesting, or any number of other natural or human factors. As a result of the "impermanence" of forests, many researchers have suggested discounting carbon credits, similar to what is proposed with leakage. A number of prominent voluntary carbon standards have now taken this approach (e.g., the Verified Carbon Standard).

As in the case of leakage, when ad hoc discounts are used, inefficiencies are created. The inefficiency is particularly problematic with permanence, however, because rental contracts, as we have discussed, provide a clear alternative. Rental contracts pay for temporary carbon storage. If forests are not permanently maintained, then rental payments would stop. As long as the buyer is liable for ensuring that the carbon credits are offset somehow, the buyer can go onto the market and buy new credits or rent new forests.

Forest Ownership

Another complexity that must be overcome to create a global program involves overlapping forest ownership. In many forests in the developed world, forests are owned privately by an individual or firm. Most carbon storage programs imagine that they must deal with only a single owner. However, even in developed economies, a great deal of forest is owned by the government or held in some type of common ownership. Here there may be many interest groups that cherish very different aspects of the same forest. A program that encourages more carbon in the forest would enhance some of those services but threaten others. For example, people who would enjoy old growth should welcome storage programs that lengthen tree rotations. However, water flows from such forests would likely be reduced as older forests tend to evaporate more water. People who like species that depend on younger forests would also be negatively affected by the carbon storage program. The carbon storage program may not be universally accepted as an improvement in forest management by these diverse interests.

In many developing economies, the issue of forest ownership is even more complex. Overlapping interests are typical in many tropical forests. The government or timber concessions may have the right to harvest the timber. But local inhabitants may have the right to harvest the wildlife, collect nontimber forest products or firewood, or graze their animals. What incentives will be given to each group to store more carbon? What if the forest is owned by a village or large family? How will the carbon program interact with the village? It is far more difficult to make transactions with villages or large families than a single forest owner. Current economic analyses have not grasped the cost of this problem at all. In principle, one would need to encourage each party to cooperate with a separate payment.

Equity

There are also important equity issues associated with forest ownership. Some of the poorest people in the world are rural inhabitants of forestlands in tropical countries. Some of the richest people in the world own forest concessions. Global forest programs may pay developing economies to store carbon on forestland, but who actually receives these payments? Do local inhabitants of these forests get any of the compensation? Is the compensation limited to timber concessions? There are important equity issues facing carbon storage programs that have not been resolved. Some of these issues may well raise the cost of the program. They will certainly raise the administrative cost. They may even dramatically affect the social desirability of the programs.

Implications of Measurement and Monitoring Limitations

The measurement and monitoring issues discussed above suggest that a project approach to collecting forest carbon has very serious limitations, particularly in the form of leakage. It may be that only broad national approaches are truly viable. Under a nationalized approached, payments would only be made for total forest carbon at the national level. Internal leakage would be offset, and payments would be made for net changes over time. International leakage would become the responsibility of the country, and the country would need to offset these if it were to receive credits. Internal issues would need to be addressed by the national authority, but failure to do so would mitigate any carbon payments. In addition, this approach would have the advantage of not requiring payment for all forest carbon, but only for positive increments over an agreed base. Broad negotiations (such as those undertaken for Indonesia) that envisage direct payments in return for broad corrective forestry practices and performance might negate the need for precise estimates of forest carbon.

Policy Conclusions

In this chapter, we have explained how carbon storage in forests has enormous potential but also that any meaningful system would be difficult to implement. Forest carbon storage could be responsible for one-quarter of all mitigation and hence cannot be ignored.

There are several important administrative and institutional hurdles that must be overcome for forest carbon storage to be effective. However, many of these hurdles have known solutions. For example, leakage and additionality are serious drawbacks to current forest projects. Universal programs can solve both problems. However, universal subsidies would involve a large income transfer to forest owners. Universal liability would involve a large income transfer away from forest owners. Some combination of liability and subsidies could provide a balanced budget approach that avoids large income transfers and provides the right incentives on the margin. Carbon sequestration and forestry are both dynamic phenomena. The carbon sequestration program must therefore be nimble with respect to time to capture these dynamics accurately. Many policy planners wish to pay only for extra carbon stored. Finally, measurement and verification are important limitations. The program must encourage least-cost measurement technology (such as LIDAR) or the administrative costs could skyrocket.

But even with these administrative innovations, there are two more issues that have yet to be addressed. The carbon storage program must be able to deal with common-property forests (forests that are owned by many). The carbon storage program must also come to terms with equity issues related to local forest inhabitants. One approach may be to nationalize the approach to internalize the leakage problem and place the ownership and equity problems with the national government, which will now have a financial incentive to address these. If carbon storage programs can overcome these administrative hurdles, there is every reason to believe forestry can live up to its mitigation potential. If the programs fail to address these issues, forestry will likely prove to be an ineffective source of carbon mitigation.

References and Suggested Readings

For more information on carbon emissions from land use, see the following:

Houghton, R. A., 2003, "Revised Estimates of the Annual Net Flux of Carbon to the Atmosphere from Changes in Land Use and Land Management 1850–2000," *Tellus Series B Chemical and Physical Meteorology,* Vol. 55B, pp. 378–390.

Intergovernmental Panel on Climate Change, 2007, *Climate Change 2007: Mitigation of Climate Change.* Contribution of Working Group III to the Fourth Assessment Report of the Intergovernmental Panel on Climate Change, 2007 (Cambridge, UK: Cambridge University Press).

Pan, Y., R. A. Birdsey, J. Fang, R. Houghton, P. E. Kauppi, W. A. Kurz, O. L. Phillips, A. Shvidenko, S. L. Lewis, J. G. Canadell, P. Ciais,

R. B. Jackson, S. W. Pacala, A. D. McGuire, S. Piao, A. Rautiainen, S. Sitch, and D. Hayes, 2011, "A Large and Persistent Carbon Sink in the World's Forests," *Science,* Vol. 333, pp. 988–993. DOI: 10.1126/science.1201609.

United Nations FAO, 2010, "Global Forest Resources Assessment 2010." FAO Forestry Paper 163 (Rome: United Nations Food and Agricultural Organization).

For more information on the cost of forest sequestration, see the following:

Richards, K., and C. Stokes, 2004, "A Review of Forest Carbon Sequestration Cost Studies: A Dozen Years of Research," *Climatic Change,* Vol. 63, pp. 1–48.

Sathaye, J., W. Makundi, L. Dale, P. Chan, and K. Andrasko, 2006, "GHG Mitigation Potential, Costs and Benefits in Global Forests," *Energy Journal,* Vol. 27, pp. 127–162.

Sohngen, B., 2010, "Forestry Carbon Sequestration," in *Smart Solutions to Climate Change: Comparing Costs and Benefits*, ed. by B. Lomborg (Cambridge, UK: Cambridge University Press), pp. 114–132.

———, and R. Mendelsohn, 2003, "An Optimal Control Model of Forest Carbon Sequestration," *American Journal of Agricultural Economics*, Vol. 85, No. 2, pp. 448–457.

For more information on designing carbon storage incentives, see the following:

Antle, J. M., S. M. Capalbo, S. Mooney, E. T. Elliot, and K. H. Paustian, 2003, "Spatial Heterogeneity, Contract Design, and the Efficiency of Carbon Sequestration Policies for Agriculture," *Journal of Environmental Economics and Management,* Vol. 46, pp. 231–250.

Marland, G., K. Fruit, and R. Sedjo, 2001, "Accounting for Sequestered Carbon: The Question of Permanence," *Environmental Science and Policy*, Vol. 4, No. 6, pp. 259–268.

Sedjo, R., and G. Marland, 2003, "Inter-Trading Permanent Emissions Credits and Rented Temporary Carbon Emissions Offsets: Some Issues and Alternatives," *Climate Policy*, Vol. 3, No. 4, pp. 435–444.

For more information on the cost of measurement and compliance, see the following:

Antinori, C., and J. Sathaye, 2007, "Assessing Transaction Costs of Project-Based Greenhouse Gas Emissions Trading," Report LBNL-57315 (Berkeley, CA: Lawrence Berkeley National Laboratory).

Macauley, M., and R. A. Sedjo, 2011, "Forests in Climate Policy: Technical, Institutional and Economic Issues in Measurement and Monitoring." *Mitigation and Adaptation Strategies for Global Change*, Vol. 16, No. 5, pp. 489–513.

For more information on programs to store carbon in forests, see the following:

U.S. Department of Agriculture, Farm Services Agency, http://www.fsa.usda.gov/FSA.

Verified Carbon Standard, www.v-c-s.org.

World Bank, 2011, Forest Carbon Partnership Facility, http://www.forestcarbonpartnership.org/fcp/.

CHAPTER

6 Mitigation and Fuel Pricing in Developing Economies*

Robert Gillingham
Independent consultant, formerly Fiscal Affairs Department, International Monetary Fund

Michael Keen
Fiscal Affairs Department, International Monetary Fund

Key Messages for Policymakers

- Low- and lower-middle-income countries contribute only about 12 percent of global carbon dioxide (CO_2) emissions, though this share is increasing (and they account for a larger share of other greenhouse gases).
- Several large, lower-middle-income countries are already important sources of CO_2 emissions; for them, much of the guidance developed in other chapters of this volume is relevant.
- Developing economies that are low emitters of CO_2 nonetheless have a critical role in finding an effective and efficient global response to the challenges from climate change: ways need to be found both to prevent "carbon leakage," as mitigation measures in high-emitting countries cause emissions to shift there, and to exploit the relatively cheap opportunities for emissions reduction there.
- Contrary to a standard mantra, an efficient approach to global mitigation does not require that developing economies charge emissions at the same rate as high-emitting countries.
- Good tax policy in these countries calls for applying to fuel use both excises (applying to both businesses and households) that at least reflect environmental damage arising at local and national levels and, on top of that, VAT or some other form of sales tax on use by final consumers.
- Fossil fuel subsidies are almost always bad policy, even apart from the increase in emissions they cause, since there are generally better ways to help the poor;

* We are grateful to David Coady, Shanta Devarajan, Philip Daniel, Stephane Kallegatte, Adele Morris, and Ian Parry for many helpful comments.

transparency and the development of social support systems are key to exiting from them.

- The problems of weak administration and poor compliance that pervade many developing economies argue for robust taxation of energy use.
- Offset schemes, whatever their intrinsic strengths and weaknesses as strategies for mitigation—which are likely to become more important in the coming years—are potentially a useful source of additional revenue for many developing countries.

This chapter considers fiscal policies toward mitigation, and those bearing on the use of fossil fuels more generally, in developing economies. It also considers, more briefly, the implications for them of mitigation measures adopted in more advanced economies. The central argument of the chapter is straightforward. The course of emissions in most, though not all, developing economies—by which we mean those in the low-income and lower-middle-income categories of the World Bank classification (see the Appendix, Table 6A.1)—is relatively immaterial to future warming. The case for them to bear substantial mitigation costs to limit emissions in the near future is correspondingly weak, even leaving aside the argument that they bear little responsibility for the cumulative stock of greenhouse gases (GHGs) and so have little duty to act. But ways need to be found both to exploit possible low-cost opportunities from mitigation there, with financial flows from developed economies having an important role, and, if need be, to avoid excessive leakage from aggressive schemes adopted in more advanced economies. Moreover, even setting climate concerns aside, there is in many developing economies—in terms simply of their own self-interest—scope for significant improvement in fiscal policies toward fossil fuels. And these will, in many cases, act in the direction of reduced emissions.

The chapter starts by looking at current and prospective contributions to global CO_2 emissions in developing economies, and then at current levels of consumption and production of fossil fuels and GHG emissions, focusing on the distribution across countries categorized by income level. The next section discusses tax design for products and activities that result in GHG emissions, recognizing both efficiency and equity concerns. This is followed by a comparison of the principles that emerge with current policies in low- and lower-middle-income countries. The next two sections address how to deal with any unwelcome equity consequences of the reforms this comparison suggests, and offset schemes, which are a source of revenue for developing economies that is currently very limited but seems likely to grow in importance.

Mitigation for the Collective Good?

Low-income and lower-middle-income countries account, collectively, for only about 12 percent of global CO_2 emissions (though a higher share of other GHGs, as seen below), and their per capita emissions are barely one-tenth of those in high-income countries (Figure 6.1). Individually, all low-income and most lower-middle-income countries have only negligible emissions. These emissions can, of course, be expected to grow, both absolutely and in percent of the total, but they will still have only a relatively modest impact on global emission totals for decades.[1] A handful of lower-middle-income countries emit significant quantities of GHGs: Five of them—India, Indonesia, Pakistan, Ukraine, and Uzbekistan—accounted for 75 percent of the total emissions of the 50 lower-middle-income countries in 2009, and 10 countries accounted for 90 percent. Efforts to curtail emissions or slow emissions growth in these countries can make a real contribution to reducing climate damage—and much the same analysis of issues and methods discussed elsewhere in this volume applies to these countries.

The focus here, however, is on low-income and the low-emitting lower-middle-income countries. What happens to their emission makes relatively little difference to atmospheric GHG concentrations. Even though many of them stand to suffer most from climate change, their inability to affect the climate outcome gives them little incentive to undertake unilaterally measures specifically to curtail emissions. To the extent that mitigation measures are

Figure 6.1. CO_2 Emissions, 2009. (Left) Global share; (right) emissions per capita (kilograms).

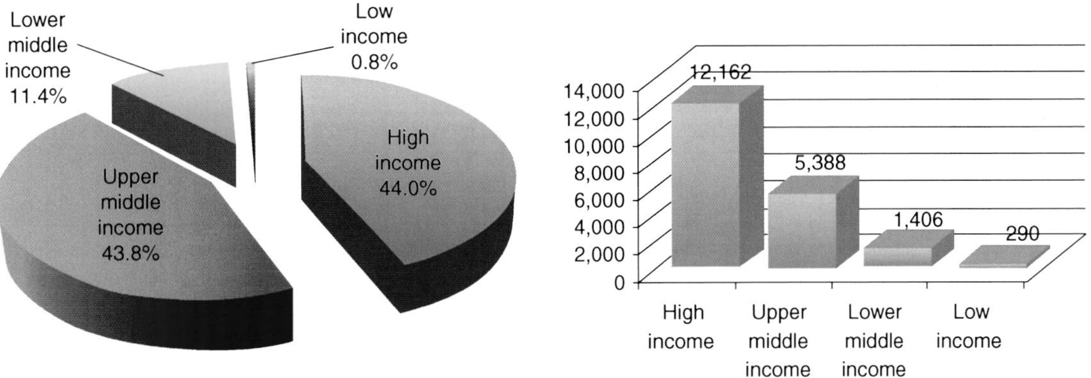

Source: Energy Information Administration (http://www.eia.gov/countries/data.cfm) and authors' estimates.

[1] Projections allowing a breakdown by income group readily comparable with these and later figures for current and past emissions do not seem to be available. For the set of countries other than Brazil, China, India, Indonesia, and Russia that are not among the advanced economies, however, the Organization for Economic Cooperation and Development projects a broadly unchanged share (at about 26 percent) of total GHG emissions from 2010 to 2050.

costly for them, the case for them to adopt, for climate reasons, aggressive mitigation strategies—in the sense of some form of carbon pricing at levels comparable to those appropriate in high-emitting countries—is correspondingly weak both in terms of their own self-interest and in terms of the wider good. There are, however, two caveats:

- Mitigation measures may be needed in these countries to avoid "leakage" from aggressive measures adopted elsewhere; that is, to avoid an increase in emissions there caused by the relocation of emitting firms or by the reduction in fossil fuel prices induced by the actions of others. This risk would be lessened if those implementing aggressive policies were to adopt border tax adjustments that impose corresponding charges on their imports from countries not imposing comparable emission prices—which, by the same token, will increase the self-interest of these countries in adopting similar measures.

- To the extent that, as will often be the case, it is cheaper to reduce emissions in these countries than in more advanced ones, efficiency (and, hence, potential gain to all) requires finding some way to realize those opportunities, quite likely involving some form of international transfer.

It may then be that pricing to reflect the global damages from GHG emissions in many developing economies need not be a priority in coordinated efforts to reduce global warming. Fiscal policies toward fossil fuels in these countries can then be driven by narrower notions of self-interest. And here, as discussed later, there is substantial scope for improvement. Before turning to this, however, it is helpful to take a closer look at patterns of fuel use and GHG emissions in these countries.

Fossil Fuel Consumption and the Distribution of CO_2 Emissions

Trends in Consumption and Emissions by Income Level

The consumption of fossil fuels is heavily concentrated in high-income and upper-middle-income countries. In 1980, these two groups accounted for 96 percent of total energy consumption (Table 6.1). This share decreased to 89 percent by 2008, with a faster-than-average rate of increase in the consumption by upper-middle-income countries more than offset by the slower-than-average growth in consumption in high-income countries (43 percent). The extremely rapid increase in energy consumption in lower-middle-income countries accounts for all of the reduction in the share of consumption in the two higher income groups, with 80 and 90 percent, respectively, of the increase in this share coming from the 5 and 10 largest emitters in the group. The shares of the bottom 45 and 40 emitters in this group were only 1.2 and 0.5 percentage points, respectively.

Table 6.1. Consumption of Energy and CO_2 Emissions, 1980–2008

	1980	1990	2000	2008	Change (Percent) 1980–2008[a]
		Share of emissions from petroleum			
World	47.9	41.7	42.6	36.7	−11.2
High income	51.9	46.9	47.4	46.0	−5.9
Upper middle income	40.6	33.7	34.7	26.1	−14.5
Lower middle income	54.1	48.5	40.5	36.6	−17.5
Low income	26.6	27.4	39.1	40.9	14.3
		Share of emissions from coal			
World	35.6	38.4	36.3	43.2	7.6
High income	31.7	34.2	31.5	31.2	−0.5
Upper middle income	41.9	44.2	44.1	57.0	15.0
Lower middle income	32.3	34.9	37.7	42.7	10.4
Low income	71.1	66.6	46.3	37.6	−33.5
		Share of emissions from natural gas			
World	16.5	19.8	21.2	20.1	3.6
High income	16.4	18.8	21.1	22.8	6.4
Upper middle income	17.5	22.0	21.2	16.9	−0.6
Lower middle income	13.6	16.6	21.8	20.7	7.1
Low income	2.4	6.0	14.6	21.6	19.2
		Total energy consumption (percent of world total)			
World	100.0	100.0	100.0	100.0	
High income	63.2	57.8	60.2	51.4	−11.8
Upper middle income	32.5	36.2	29.6	37.3	4.8
Lower middle income	3.5	5.2	9.5	10.6	7.0
Low income	0.8	0.8	0.7	0.8	−0.1
		CO_2 emissions (percent of world total)			
World	100.0	100.0	100.0	100.0	
High income	61.0	54.3	56.9	46.7	−14.3
Upper middle income	34.3	39.0	32.0	41.3	7.1
Lower middle income	3.9	5.8	10.4	11.3	7.4
Low income	0.9	0.9	0.7	0.7	−0.2
Memorandum items:					
Total energy consumption (quadrillion BTUs)	281.8	345.4	393.4	488.0	73.2
Total CO_2 emissions (billion metric tons)	18.3	21.5	23.5	30.1	64.0

Source: Energy Information Administration (http://www.eia.gov/countries/data.cfm) and authors' estimates.
[a] Percentage change for level and percentage point change for share.

By any metric, low-income countries have not been significant contributors to CO_2 emissions.[2] Indeed their share of CO_2 emissions has even decreased, even though their share of energy consumption has remained constant. This result stems—at least in part—from the fact that the income groupings are defined based on current income levels. A number of countries that have grown rapidly—for instance, China and India—have moved into middle-income categories, increasing the rate of growth in energy consumption and CO_2 emissions in these categories and reducing the rate of growth in the low-income category. Again, the few large consumers/emitters among lower-middle-income countries accounted for almost all of the 7.4 percentage point increase in the share of this group.

The rate of growth in emissions largely mirrors the growth in energy consumption. However, the overall rate of emissions growth is slower, reflecting an average reduction in emissions per unit of energy consumed. This reflects primarily a shift in the composition of energy sources toward fuels that are less carbon intensive on average. Globally, there has been a shift away from petroleum and toward coal, natural gas, and other nonemitting energy sources such as nuclear power and renewables (Table 6A.2, in the Appendix), with a concomitant shift in emissions sources (Table 6.2). The result has been a net increase in emissions intensity (Table 6.3), although the shift to coal, which has the highest emissions intensity, partly offset the gains from the shifts to natural gas and renewables. High-income countries have shifted from petroleum and coal into natural gas and other energy sources, while middle-income countries, especially upper-middle-income countries because of China, have shifted into coal. Low-income countries, on the other hand, have seen an extreme shift out of coal into primarily natural gas, most likely due to the increased availability of this energy source.

In short:

- Low-income and the vast majority of lower-middle-income countries have contributed very little to either the level or the growth in global emissions, primarily because slow growth has constrained the growth in their energy consumption.
- In addition, low-income countries—but not lower-middle-income countries in general—have shifted away from coal and toward cleaner hydrocarbon energy sources, again mitigating their contribution to CO_2 emissions.

Stringent mitigation measures by low-income countries and most lower-middle-income countries are, thus, not a necessary condition for reducing global CO_2 emissions.

[2] Low-income countries, for instance, have accounted for about 2 percent of cumulative CO_2 emissions since 1850.

Table 6.2. Distribution of Emissions Sources, 1980–2008

	1980	1990	2000	2008	Change (Percent) 1980–2008[a]
		Share of emissions from petroleum			
World	47.9	41.7	42.6	36.7	–11.2
High income	51.9	46.9	47.4	46.0	–5.9
Upper middle income	40.6	33.7	34.7	26.1	–14.5
Lower middle income	54.1	48.5	40.5	36.6	–17.5
Low income	26.6	27.4	39.1	40.9	14.3
		Share of emissions from coal			
World	35.6	38.4	36.3	43.2	7.6
High income	31.7	34.2	31.5	31.2	–0.5
Upper middle income	41.9	44.2	44.1	57.0	15.0
Lower middle income	32.3	34.9	37.7	42.7	10.4
Low income	71.1	66.6	46.3	37.6	–33.5
		Share of emissions from natural gas			
World	16.5	19.8	21.2	20.1	3.6
High income	16.4	18.8	21.1	22.8	6.4
Upper middle income	17.5	22.0	21.2	16.9	– 0.6
Lower middle income	13.6	16.6	21.8	20.7	7.1
Low income	2.4	6.0	14.6	21.6	19.2

Source: Energy Information Administration (http://www.eia.gov/countries/data.cfm) and authors' estimates.
[a] Percentage change for consumption and percentage point change for share.

Trends in Energy Efficiency

A somewhat different picture emerges if one looks instead at the efficiency with which low- and middle-income countries use energy, both in terms of emissions and energy consumption. This picture is, however, somewhat different depending on how GDP is compared across countries.

Using market exchange rates (in the upper part of Table 6.3), the worldwide carbon intensity of production fell by 12 percent between 1980 and 2008. This stemmed entirely from improved efficiency in high- and low-income countries, with a very small contribution from low-income countries to the aggregate. Moreover, almost all of the increase in efficiency has stemmed from a reduction in the quantity of energy used to produce output. Energy efficiency fell for middle-income countries, as evidenced by the increase in Btu[3] per U.S. dollar of GDP for both subgroups. Measuring GDP in purchasing power parity terms (lower part of Table 6.3)—which adjusts, for

[3] Btu stands for the British thermal unit, which is a measure of energy expended.

Table 6.3. Emissions and Energy Efficiency, 1980–2008

	1980	1990	2000	2008	Change (Percent) 1980–2008[a]
	Kilogram of carbon per thousand 2005 U.S. dollars of GDP (exchange rates)				
World	688	705	596	603	–12.4
High income	595	466	418	372	–37.5
Upper middle income	1,225	1,890	1,327	1,346	9.9
Lower middle income	989	1,258	1,506	1,300	31.4
Low income	1,334	1,281	900	739	–44.6
	Btus per thousand 2005 U.S. dollars of GDP (exchange rates)				
World	10,723	11,384	9,965	9,785	–8.8
High income	9,669	8,046	7,406	6,643	–31.3
Upper middle income	16,858	28,230	20,522	19,720	17.0
Lower middle income	13,680	17,935	22,873	19,678	43.8
Low income	19,596	18,243	15,002	12,999	–33.7
	Kilogram of carbon per thousand 2005 PPP dollars of GDP				
World	597	589	490	462	–22.6
High income	626	488	434	383	–38.9
Upper middle income	639	920	647	637	–0.3
Lower middle income	348	436	519	449	28.9
Low income	327	308	200	157	–52.1
	Btus per thousand 2005 PPP dollars of GDP				
World	9,301	9,512	8,190	7,495	–19.4
High income	10,175	8,426	7,687	6,838	–32.8
Upper middle income	8,793	13,737	10,006	9,331	6.1
Lower middle income	4,814	6,211	7,884	6,791	41.1
Low income	4,796	4,389	3,338	2,753	–42.6

Source: Energy Information Administration (http://www.eia.gov/countries/data.cfm) and authors' estimates.
Note: Btu = British thermal unit; PPP = purchasing power parity.
[a] Percentage change for consumption and percentage point change for share.

example, for labor-intensive nontradable services being cheaper in poorer countries—generally leads to higher GDP numbers and therefore to lower emission and energy intensity.[4] Now (almost) all of the increase in energy and emissions efficiency occurs in high-income countries, but the level of efficiency in other income groups seems, relative to those countries, much higher. The broad qualitative pattern of changes is much the same, except that emissions intensity in upper-middle-income countries falls on a purchasing power parity basis, but rises when market exchange rates are used.

[4] Efficiency levels for high-income countries are not greatly affected since the United States is taken as a benchmark in the purchasing power parity calculations.

The lesson here is that even though low-income and many lower-middle-income countries account for a very small share of global energy use and global emissions, they have room for increases in efficiency that will be of value if and when these countries achieve sustained growth—and which the offset schemes that we will now discuss attempt to exploit.

Other Greenhouse Gases

Methane and nitrous oxide are the next two most important anthropogenic GHGs after CO_2. The levels of these GHGs, measured in CO_2 equivalent, are significant, although less in aggregate than CO_2 emissions (Table 6.4). Low- and lower-middle-income countries account for a much larger share of these emissions. Discounting the incomplete data for 2005, the share of emissions in low-income countries is on the order of 10 percent for the two GHGs combined, while lower-middle-income countries account for roughly one-quarter.

Agriculture and livestock cultivation are the most important sources of anthropogenic methane and nitrous oxide. Consequently, low- and lower-middle-income countries could become a larger source of these GHGs as, for instance, widespread use of nitrogen-based fertilizer is adopted in more poor countries. These emissions should, in principle, be subject to pricing that reflects their contribution to the buildup of GHGs. In practice, however, they do not lend themselves to the same methods of taxation

Table 6.4. Combined Methane and Nitrous Oxide Emissions, 1990–2005

	1990	1995	2000	2005[a]
	(Millions of metric tons of CO_2 equivalent)			
World	7,891	8,126	8,627	7,351
High income	2,532	2,500	2,370	2,291
Upper middle income	3,043	3,206	3,389	2,170
Lower middle income	1,698	1,772	2,070	1,639
Low income	618	648	798	1,251
	(Percent of world total)			
World	100.0	100.0	100.0	100.0
High income	32.1	30.8	27.5	31.2
Upper middle income	38.6	39.4	39.3	29.5
Lower middle income	21.5	21.8	24.0	22.3
Low income	7.8	8.0	9.2	17.0

Source: World Resources Institute (http://earthtrends.wri.org/) and authors' estimates.

[a] Data for 2005 are fragmentary. We assume growth rate for missing countries is equal to the average growth rate for countries in the same income class for which data are available.

or cap-and-trade that are the focus of this volume and chapter, lacking an analogue to the possibility for relatively upstream monitoring of the use of fossil fuels to which CO_2 emissions are very mechanically related. For the foreseeable future, the primary mitigation instrument in this area, at least in lower-income countries, is thus likely to be support for efficient agricultural and husbandry techniques.

Tax Design Considerations

Low-emitting developing economies face two key questions: Should they nonetheless price carbon emissions and, if so, at what level? How do other considerations affect how they should tax fossil fuel use?

Should Energy Taxes in Developing Economies Reflect Climate Concerns?

It is a standard mantra that a globally efficient approach to mitigation requires that all countries apply the same charge to CO_2 emissions. Even leaving aside the practical consideration above that emissions from many developing economies are tiny, this mantra is incorrect as a matter of principle for two reasons.

First, the mantra applies only if issues of income distribution *among* countries are either immaterial or adequately addressed by transfers from richer countries to poorer. Neither of these conditions seems a reasonable assumption on which to base policy. And when it fails, a strong case can be made on equity grounds—and this is quite independent of the "responsibility" argument—for taxing emission at a lower rate in poorer countries. Indeed, given the low level of their emissions noted above, a carbon charge of zero—or at least low enough to avoid significant leakage—might well be reasonable on these grounds.

Second, the mantra ignores the interaction of CO_2 emissions charges with the rest of the tax system (and, potentially, other market imperfections, too, such as local pollution, discussed below). The implications of this can be complex. In some cases, it may point to corrective charges on CO_2 emissions being *lower* than would otherwise be the case because the good they do in correcting environmental damage is offset, in part at least, by the damage they do to the government's wider revenue-raising objective by reducing the level of economic activity. In other cases, however, these interactions may point to taxation at a *higher* rate than otherwise—perhaps as a response to informality, as discussed further below. The key point here is that carbon charges need to be seen as part of countries' wider revenue-raising systems and in light of country-specific challenges in these areas. It is to these that we turn next.

Climate Aside, How Should Fossil Fuels Be Taxed in Developing Economies?

The same general principles apply as in developed economies, as discussed elsewhere in this volume—but with some potentially important differences in application.

Looking beyond climate damage, the use of fossil fuels may be associated with adverse external effects at the local or national level that policymakers can and should address. These might include, for instance, the congestion associated with vehicle use, local noise pollution, or SO_2, NO_x, and particulate emissions. Such externalities can call for the use of "corrective" taxes—intended to change behavior rather than to raise revenue. To avoid unintended distortions, these should be targeted as closely as possible to the source of the externality. So it is only carbon-related externalities that call for correction in the form of carbon charges. Congestion, for instance, is better handled by explicit congestion charging (per-mile tolls rising and falling during the course of the rush hour), as is beginning in some cities in the developing and emerging world. (Indeed, over the longer term, congestion pricing is a promising source of own finance for lower-level governments.) The potential to fine-tune corrective taxes in this way, however, is constrained both by limited administrative capacity in many developing economies, which may have higher priorities, and by limited knowledge of the magnitude of the relevant external costs in developing economies. A broad-based carbon tax could go a long way to addressing these local externalities; early work for Pakistan, for instance, suggested that a carbon tax in the order of US$25 per tonne of CO_2 would be more than warranted solely in these terms. Failing that, fuel taxes, in particular, are likely to retain for some time a potentially important role as proxying corrective charges on congestion and other underlying sources of harm associated with vehicle use.

Importantly, such charges intended to correct for externalities associated with fuel use should be charged at the same rate (in the absence of good reasons to do otherwise) on use by both businesses and final consumers—the damage is the same in either case. There should be no exemption, for instance, for fuels used in road haulage, and congestion charges, if applied, should not be remitted for business travel. So far as is practicable, the rates applied to different petroleum products should differ to reflect differences in the externalities generated; motor fuel taxes, for instance, should reflect congestion and accident externalities (which are relatively large), while fuel oil for power generators or oil for home heating should not.

Externalities aside, taxes on energy use have a role as one among many devices for meeting a government's revenue needs. If the government can effectively implement a full range of taxes on consumption (a big "if" that is returned to below), then this revenue-raising motive requires (in contrast to the environmental motive) that only use by final consumers be taxed, not use as a business input. The reason for this is that taxes on input used by businesses will typically lead them to make choices (their mix of inputs and degree of vertical integration of their activities) that differ from those that would be driven by the underlying prices that matter for social costs and benefits. So taxing intermediate use effectively reduces the aggregate level of output and thereby makes raising whatever revenue is needed more costly for society. Thus, looking only to revenue considerations, fuels should typically be subject to whatever tax a country levies on final sales; in practice, this typically means the value-added tax (VAT), which effectively excludes business use from taxation by giving registered businesses a credit or refund of taxes charged on their inputs. So, for instance, VAT should apply to all electricity consumption (residential, industrial, and commercial), but credits or refunds would be provided to industrial and commercial users to ensure that the tax really only bites for residential use.

When externalities are present *and* revenue is a concern, these considerations point to two distinct charges on fuel use: An excise (in "specific" tax form—that is, as a fixed monetary amount—since the damage does not itself vary with the price of the good) and a charge imposed—on the price including the excise[5]—as part of the wider VAT system. The excise itself might, in principle, be somewhat lower or higher than the marginal environmental harm, as noted above, to best take account of interactions with the wider tax system—but in practice, this may be a subsidiary consideration.

The question then is what rates of taxation should be applied to final consumption of fuels. The first general principle here is that items of final consumption should be taxed at a higher rate the more their use is associated with "leisure" (in the sense of untaxed activity)—to limit the overall distortionary impact of the tax system, a condition that is closely associated with the familiar "inverse elasticity rule"; that is, that tax rates should be higher on goods in more inelastic demand. The second general principle is that the tax rate should be higher, all else equal, the more its consumption is concentrated among the better off in order to meet equity goals.

There are some insights from this that are of fairly general applicability. Domestic transport fuel, for instance, is likely to be a candidate for relatively heavy taxation on the first efficiency criterion. But the full implications will depend on each country's own circumstances; it matters for instance, whether individuals use fuel mainly for leisure or mainly for work-related travel. And these, moreover, are often issues on which empirical evidence is weak, and

[5] This sequencing—VAT charged on the price including the excise—ensures that changes in the general rate of VAT do not discriminate across goods in the sense that they do not affect their relative prices.

hence, so too is the case for charging differential tax rates (even leaving aside the implementation costs of doing so, which can be considerable). On the second, equity criterion, one general lesson is that differentiating rates of taxation across items of final consumption is rarely the best way of pursuing distributional objectives. This is most clearly the case in more advanced economies, with a range of income-related and other instruments available to help the poor. Even in lower income countries, however, the poor may benefit more from targeted health and education spending, tax subsidies for work, or the development of simple social safety nets (as discussed below) than they would from reduced tax rates on particular items. And indeed, the rationale offered for reduced rates—or, as discussed later, outright subsidies—is often more fundamentally flawed. For example, some developing countries have sought to address concerns that taxes on household electricity use, in particular, would have adverse distributional effects by introducing a reduced VAT rate for low levels of usage; but, as with fuel subsidies more generally, the concern is often misplaced—the poorest households in many developing countries, for instance, are unlikely to have any access to electricity.

In practice, the case in principle for applying differential rates of VAT to fuel products is, thus, often weak. Add to this the administrative and compliance problems that such rate differentiation creates—likely to be especially evident in lower-income countries, and a presumption emerges in favor of applying a single rate of VAT.

The challenges of revenue administration in many developing economies may also be important in setting the level of excises on petroleum fuels in particular and fossil fuels in general. They likely tilt the balance still further in favor of relatively heavy taxation of energy use in general and hydrocarbons in particular, for two reasons:

- First, fuel taxes—the excises in particular—are, or should be, among the easiest taxes to administer and with which to comply. Charges can be applied at mine mouth, import, or refinery gate—meaning there are relatively few points that the authorities need to control (see Chapter 2). National oil companies can also provide a natural point of control. The relative ease of taxing fuels compared with, for instance, personal incomes, argues for greater reliance on this source of revenue than would otherwise be the case.

- Second, the implication of the large informal sectors—businesses that are less than fully tax compliant—in many developing countries is that the government cannot tax households' final consumption or businesses' profits as it would like. In this case, there is a strong case for taxing that consumption and those profits indirectly by taxing inputs used in their production. And for many enterprises, energy is one such input. Taxes on fuel use then have particular appeal in substituting for missing taxes on the sales and profits of noncompliant enterprises.

However, these arguments should not be pushed too far. Heavy energy excises also bear on compliant firms,[6] for instance, and labor market distortions can amplify potentially adverse effects on employment (as has been identified as a concern in South Africa, for instance). They do, nonetheless, reinforce the case for vigorous taxation of energy use in developing economies—combined, if need be, with measures to protect the poorest along lines discussed below.

Pricing of Hydrocarbon Products in Low- and Lower-Middle-Income Countries

Clearly, developing economies, like others, should set the prices of hydrocarbon products to consumers and businesses at least at the levels that the tax policy criteria discussed previously imply would be appropriate in the absence of the GHG global externality. Addressing climate change would then require setting those prices still higher, to an extent reflecting considerations also raised above. Abstracting from this latter concern, this section asks whether developing economies already set tax rates at or above the levels that would be in their own self-interest, in the absence of climate concerns, for efficient revenue mobilization and to address local externalities.

Gasoline and Diesel Prices

There is no comprehensive database on hydrocarbon prices similar to the data on energy production and consumption and GHG emissions used above. However, GIZ, the German agency for international development, does collect the prices (including taxes) of gasoline and diesel in over 170 countries biennially (see Table 6A.3 for a summary of trends). Based on these data, there is little evidence that low-income countries significantly subsidize gasoline and diesel. The average price of gasoline in low-income countries in December 2010 was US$1.26 per liter. Prices ranged from US$0.80 per liter in Myanmar to US$1.71 in the Central African Republic and Malawi (Figure 6.2). The average price was slightly lower for countries that export petroleum.

[6] In principle, a higher VAT rate on energy use might for this reason be preferable to a higher excise: It would have the same effect on firms not complying with their VAT obligations, but not for the compliant (which would be entitled to credit or refund of the tax). Differential VAT rates can, though, add considerably to the difficulties of implementing and complying with a VAT, so that few countries do this in practice.

Figure 6.2. Gasoline Prices in Low-Income Countries, December 2010
(U.S. cents per liter, oil exporters in red)

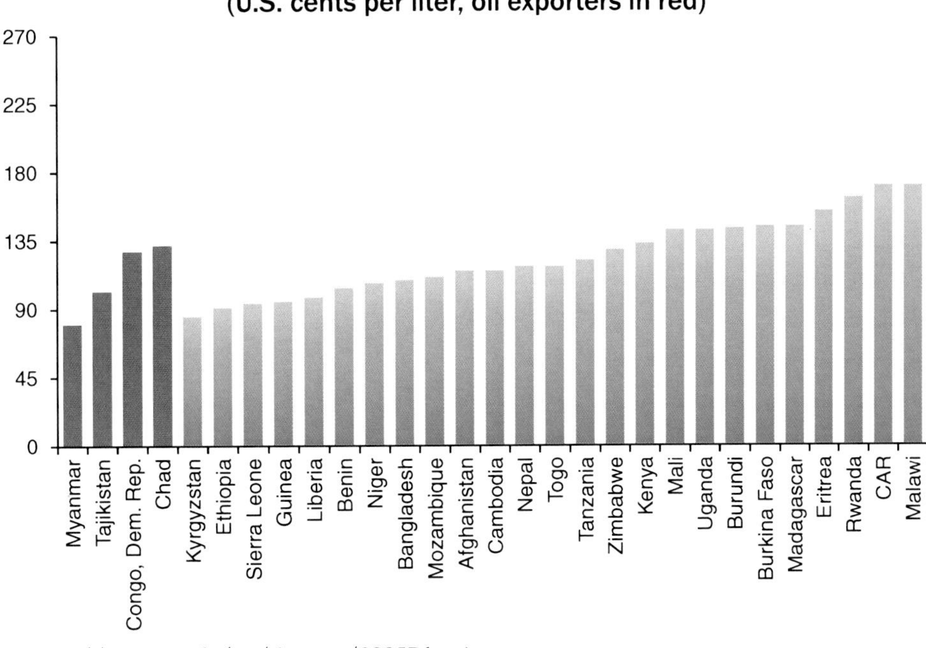

Source: GIZ (http://www.gtz.de/en/themen/29957.htm).

The tendency for oil exporters to place a low price on or even subsidize petroleum products is more pronounced for lower-middle-income countries (Figure 6.3): within this group, the average price for oil exporters in December 2010 was US$0.28 per liter lower than for nonexporters. This difference cannot be completely explained by transportation costs since a number of exporters of crude oil have no or insufficient refining capacity and must import at least some of the petroleum products they use. Even nonexporters in the lower-middle-income group had prices lower than the global average, and their overall average was the lowest among the four groups (although oil-exporting countries in the upper-middle-income category had a lower average than exporters in the lower-middle-income category).

For completeness, the distribution of gasoline prices among upper-middle-income and high-income countries is presented in the Appendix (Figures 6A.1 and 6A.2). The range in prices is very wide, with high-income European countries—including those that export oil—at the high end. Other oil-exporting countries in these categories tend to be at the low end of the distribution.

Prices of Other Hydrocarbon Fuels

As noted, the data on the pricing of other hydrocarbon fuels are less comprehensive. For natural gas, what subsidies persist appear to be in the former Soviet Union and, perhaps, in domestic markets of some other producers; low-income countries are primarily importers. More generally, and

Figure 6.3. Gasoline Prices in Lower-Middle-Income Countries, December 2010 (U.S. cents per liter, oil exporters in red)

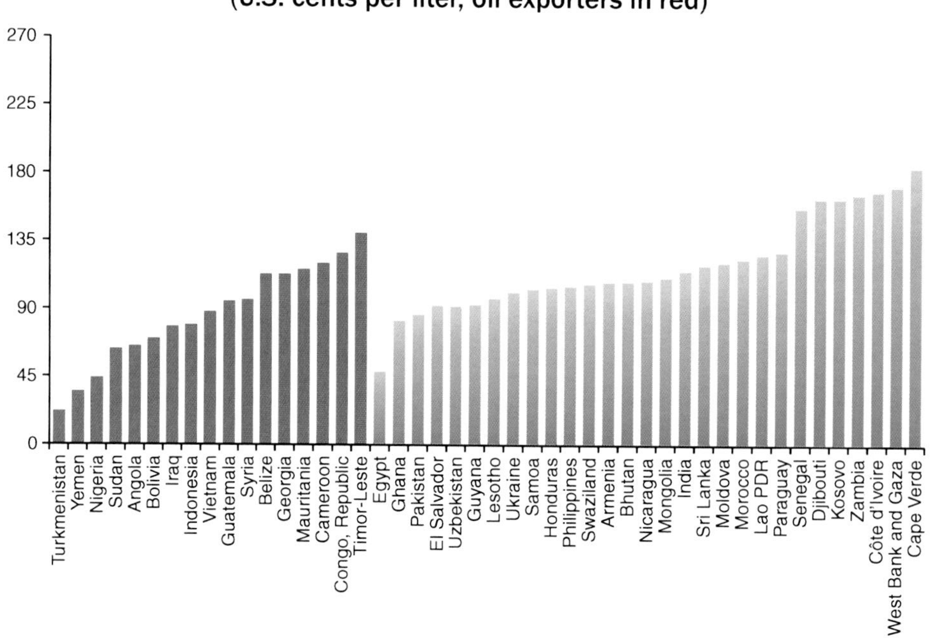

Source: GIZ (http://www.gtz.de/en/themen/29957.htm).

especially for fuels used to generate electricity, the problem seems to be less with the pricing of the fuel than with the pricing of the electricity.

Electricity and Public Transportation Subsidies

Direct price subsidies for coal—and, for that matter, subsidies for other fuels used to generate electricity—are less of a problem than the subsidies on electricity itself. For instance, *Business Insider* (May 25, 2011) reported that in China:

> The State Grid Corp of China warned [in May 2011] that it would shut down some generators over the summer, reducing output by as much as 40 gigawatts, according to *Global Times*. Power companies are struggling as coal prices have risen 75 percent since 2007, while electricity prices have risen only 15 percent. China's five biggest coal power businesses have posted combined losses of $9.23 billion since 2008. Energy analyst Li Chaolin tells *Global Times:* "Many coal plants have shut down their generators because the more they produce, the bigger the losses they will suffer."
>
> An editorial in the Oriental Times complains: "The coal industry is in a market economy, but the power industry is still in a planned economy."

A number of low-income countries subsidize electricity, creating significant budgetary problems. The electricity is often produced by a public enterprise, and the subsidies can remain hidden until they must be recapitalized. Several countries have had electricity subsidies costing more than 2 to 3 percent of their GDP.

Subsidies for public transportation can also contribute to the overuse of petroleum products even in the absence of mispricing of gasoline and diesel. As with electricity, countries often subsidize public transportation by allowing public enterprises to endure significant losses on their operations. Some studies for advanced economies have found high operating subsidies for mass transit in major cities to be warranted by various scale economies and the reduction of auto externalities (which are high in congested cities). Where the balance of considerations lies in pricing public transport needs assessment on a case-by-case basis.

Addressing Unwelcome Consequences of Subsidy Reform

Reforming Subsidies

The benefits of universal fuel subsidies commonly accrue, in practice, to the highest income groups—making them an extremely costly approach to protecting the welfare of poor households. For example, taking the poorest 20 percent of households as the target "poor" group, the cost to the budget of transferring one dollar to this group via gasoline subsidies is commonly found to be about $33 (i.e., $1/0.03). This reflects the fact that over 97 out of every $100 of gasoline subsidy "leaks" to the top four quintiles. Even for kerosene, which is widely used for heating and cooking in poor households, this (budgetary) cost-benefit ratio is generally around $5 (i.e., $1/0.19).

Such high leakage of subsidy benefits means that there is likely to be a high return to developing more effective ways of protecting the real incomes of poor households. For example, if $15 out of every $100 allocated to a safety net program is absorbed by administrative costs and 80 percent of the remaining $85 in beneficiary transfers reaches the poor (or 68 percent of the total budget), then the (budgetary) cost-benefit ratio for such a program is $1.5 (i.e., $1/0.68), which is substantially lower even than for kerosene subsidies.[7] In addition, the extent of protection that can be given to the poor via kerosene subsidies without severely disrupting fuel markets is very limited. Relatively low kerosene prices result in substitution of kerosene for diesel (legally or illegally) and often lead to shortages for rural households and smuggling to neighboring countries with higher prices.

[7] A full comparison between price subsidies and alternative ways of supporting the poor would also consider their incentive impacts and those of the tools used to finance them.

However, since eliminating fuel subsidies can still have a sizeable adverse impact on poor households, reform strategies should include measures to address equity concerns. Where an effective social safety net exists, expanding the budget for these programs can address concerns for poverty while containing the fiscal cost. For countries that do not have access to effective safety net programs, a more gradual reform approach is desirable if fiscal conditions allow. This could involve maintaining kerosene subsidies over the short term and using existing programs that can be expanded quickly, possibly with some improvements in targeting effectiveness (for instance, school meals, reduced education and health user fees, cash transfers to vulnerable groups, or subsidies for consumption of water below a specified threshold).

Similarly, other public expenditures, such as on education and health expenditures, as well as infrastructure expenditures such as roads and electrification schemes, could be expanded.

Increasing retail prices to reduce fuel subsidies is always politically difficult. At the most fundamental level, support for such subsidies can indicate a lack of faith in government: In oil-exporting countries in particular, there can be a sense that they are a way for the citizenry to enjoy benefits from their common resources in a way that sidesteps empowering government. Dealing with subsidies can be a key adjunct of wider governance reform.

An effective and credible public information campaign can be a key (indeed likely indispensible) way to increase public support for price increases by informing the potential beneficiaries (consumers and taxpayers) about the drawbacks of existing subsidies and the benefits of reform. This could, in particular:

- *Highlight the fact that subsidies provide incentives for inefficiently high levels of fuel consumption and that the associated fiscal costs can be detrimental to growth and poverty reduction.* Eliminating subsidies will encourage more efficient energy consumption and thus reduce the impact of future international price increases on the economy. In addition, subsidy reform will contribute to fiscal sustainability and economic growth, which are crucial for sustained poverty reduction. A growing economy will also enhance households' capacity to absorb price shocks for key commodities and can be the most effective approach to limiting the adverse impact of general price increases.

- *Highlight the fact that higher income groups benefit the most from fuel subsidies and that neighboring countries with higher prices are often substantial beneficiaries through cross-border smuggling.* The evidence clearly shows that most of the benefit from lower fuel prices goes to higher income groups. When relevant, governments should also highlight that subsidies promote smuggling, shortages, black market activities, and corruption.

- *Highlight the fact that retail price changes reflect fluctuations in international prices beyond the government's control. All countries face price fluctuations and need to adjust to this reality.* Passing higher international prices through to domestic prices provides the appropriate incentive to consumers to reduce fuel consumption and thus limits the adverse impact on the economy as a whole. If the domestic prices of exporters do not adjust, they bear the opportunity cost of reduced exports and the efficiency costs of mispricing.

- *Educate the population about the importance of fuel tax revenues in financing priority public expenditures.* This should highlight the importance of revenues for financing a range of high-priority public expenditures such as improvements in education, health, and physical infrastructure. During times of relatively high price increases, it can help to clearly identify the decreases in other priority expenditures that would have to be made if subsidies increase. Not least, it is essential to transparently record subsidies on-budget to ensure that they have to compete with these other sectors for available financing.

Avoiding the recurrence of fuel subsidies requires a new approach to fuel pricing in many countries. In countries with fuel subsidies, the government's control of domestic prices creates the impression that price changes simply reflect government policy, rather than international factors, with political pressure to avoid passing through increases in international prices but, on the other hand, to pass through decreases. The need for effective communication is neverending. Preventing the recurrence of subsidies can be just as difficult as eliminating them in the first place until the public becomes used to the implications of international prices for domestic prices.

Improving Pricing Mechanisms

The first best approach to petroleum pricing is to implement a fully liberalized regimen, accompanied by appropriate regulation to ensure competition. As an interim measure, however, governments can adopt automatic pricing mechanisms. But this cannot be relied on to solve the problem. Many countries have abandoned such mechanisms, or abandoned subsidy reform programs, in times of sharp increases in international prices. For example, Ghana adopted an automatic mechanism in January 2002 only to abandon it in January 2003. The increasing fiscal cost of incomplete pass-through led to the reinstatement of the mechanism in February 2005, but it was again abandoned in early 2008 when international prices increased sharply, and domestic prices remained fixed from May to November 2008. Similarly, Indonesia began a subsidy reform in 2005 with the intention of eliminating subsidies and fully passing through increases in international prices. However, this policy was abandoned by the end of 2007. So both governments that control prices and those with automatic pricing mechanisms have struggled to fully pass-through prices during periods of sharp increases in international prices. Consistent with this, median pass-through exceeded 100 percent at the end of 2006, but decreased substantially thereafter.

The fragility of automatic price adjustment mechanisms often reflects the reluctance of governments to fully pass-through sharp international price increases that they believe may be temporary. If, however, such price increases are persistent, this "wait-and-see" approach can result in escalating subsidies, and substantial increases in domestic fuel prices are eventually required. Since the public is likely to be more concerned about large price increases, reform becomes more difficult, and subsidies become entrenched. To make automatic pricing adjustments more attractive, smoothing mechanisms can be incorporated. These smoothing rules can: lessen, in the short term, the magnitude of retail price changes compared to full pass-through; ensure full pass-through of price changes over the medium term; and avoid long periods of fixed prices that eventually necessitate large retail price increases if international price increases turn out to be persistent.

Especially where smoothing rules are not in place, pressures can arise to change tax rates on fuels so as to avoid large movements in energy prices. Even leaving aside the difficulty of distinguishing temporary from permanent price shocks, the issue is a complex one. Ad valorem taxes, such as the VAT (taxes charged as a proportion of the price), actually amplify the underlying effects of fuel price movements on final prices to consumers: With a 20 percent VAT, for instance, a $1.00 increase in underlying fuel prices translates into a $1.20 increase in fuel prices at the pump. Whether this amplification is appropriate is by no means clear. Box 6.1 looks at these issues in more detail and suggests that an appropriate response—again, environmental considerations aside—may be to adjust taxes on fuels in response to a permanent increase in oil prices in such a way that total tax falls as a percent of the fuel price but rises as a monetary amount per gallon.

Box 6.1. How Should Taxes Vary with Oil Prices?

To see the rationale for the strategy set out in the text, suppose first that it was decided in response to an increase in oil prices to hold the tax constant as a monetary amount per unit (which would mean reducing any ad valorem component). In this case, with a reduction in demand consequent upon the increase in price, tax revenue will fall. But this is unlikely to be optimal; and to recover this revenue loss, at least partly, means raising tax rates on all commodities—including fuels. So an optimal response likely involves increasing the monetary amount of the tax per unit.

But can it be optimal to increase the monetary amount by so much as to leave it unchanged as a proportion of the oil price? Almost certainly not, because then tax revenue would increase (assuming, as is reasonable, that the elasticity of demand is less than unity)—and that too cannot be optimal if the original level of revenue was seen as appropriate. So the tax should increase by less than this amount.

For the environmentally motivated element of the charge, the question is whether the change in total demand induced by a change in oil prices materially affects the marginal environmental damage from oil use that this element is intended to reflect. In many cases it will not, in which case, no change is needed on these grounds.

Source: Authors.

Promoting Transparency

Fuel subsidies are often difficult to measure and evaluate, partly because of definitional and measurement issues. In some cases, fuel subsidies do not appear in the fiscal accounts at all, making them difficult to track, quantify, and assess. And even when they are reflected in the fiscal accounts, they can be hidden in broader aggregates. The inability to identify properly the costs of petroleum product subsidies, and the winners and losers they create, hinders efforts to assess them properly and, hence, the ability to undertake reforms.

Fuel subsidies should be recorded transparently in government accounts. Where appropriate, they should be recorded in the budget and explicitly identified. Off-budget subsidies should be identified and recorded in separate accounts. This may require improvements in the budget classification systems. Arrangements whereby international or national oil companies provide subsidies to consumers without explicit budget support should be clearly defined and described in budget documents.

Transparency is especially important for oil exporters where the opportunity cost of fuel subsidies is the revenue foregone by not charging international prices domestically. Oil producers that record fuel subsidies explicitly in the budget include Indonesia, Iran, Malaysia, Sudan, and Yemen. Some countries have implemented specific subsidy reporting systems designed to help raise public awareness.

Financial Flows (and Other Effects) from Developed Economies

Sometimes overlooked is the possibility that developing economies can attract finance from abroad to finance mitigation and other climate-related actions. There are three potential sources of revenue:

- First, some have argued for a global cap-and-trade scheme, with allocations of emissions rights to developing economies sufficient to ensure that they would be net sellers to other countries.
- Second (as discussed in Chapter 7), under the Copenhagen Accord, developed economies have committed to provide finance to developing economies to help meet costs of adaptation and mitigation "in the context of meaningful mitigation actions," suggesting that some effort is indeed expected of developing economies themselves.
- The third potential source is from offset schemes—provided for in the Kyoto Protocol, and no doubt to be included in other and successor agreements—by which public and private entities in developed economies facing binding emissions commitments can meet those obligations in part by investing in projects in developing economies that are certified as implying a corresponding emissions reduction there. The primary vehicle for this is the clean development mechanism (CDM), although there are other possibilities too, including voluntary schemes. Such schemes are attractive to investors to the extent that the costs of emission reduction in developing economies are lower than those in developed economies. This can often be the case—given both the use of relatively dirty technologies in developing economies and the increasing difficulty of cutting emissions in developed economies —as requirements become more demanding.

Such schemes can yield revenue to governments in developing economies in two ways, although these revenues are, to some degree, offset by the costs of mitigation. They are a direct source of revenue when government entities operate the emissions reducing project. And they can be a source of tax revenue from private projects, too, in so far as tax is chargeable on various aspects of such projects. This latter point raises a range of detailed issues as to how the host country should treat these projects for tax purposes—whether VAT and/or stamp duties should be charged on the transfer of credits, for instance, and how to tax the profits associated with such projects. Practice seems to vary: Some countries offer preferential tax treatment for such schemes, while others do not. But the appropriate policy line seems clear: Except to the extent that such projects generate local externalities of the kind stressed above, there is no particular case for taxing them differently from any others.

The sums raised by CDM projects have so far, however, been relatively limited and have accrued to just a few of the larger developing economies. To a large extent, this reflects recognized limitations of the CDM process, including high transaction costs associated with the verification that emissions reductions are indeed additional (that is, would not have occurred without the supported project). The problem of establishing the counterfactual is a serious difficulty with offset schemes: What would have happened in their absence? This is a key reason why avoided deforestation is not covered by the CDM. Arguably, the counterfactual issue renders the offset approach problematic as an approach to efficient mitigation. What matters for present purposes, however, is simply that offset schemes offer some potential for welfare gains in developing economies. As the CDM process becomes smoother and more encompassing in the coming years—and the need to make it so is widely recognized—the opportunities that offset schemes present for developing economies will, it is to be hoped, become more evident.

As yet little noted, however, is the potentially adverse impact of aggressive carbon pricing in large emitters on the increasing number of lower-income oil exporters. By reducing demand for and hence the world price of oil, such policies would reduce the resource rents that these countries can enjoy. There has, as yet, been little recognition of this issue, let alone any discussion of possible measures to reduce the impact.

Conclusion

Given the importance of low-emitting developing economies to the climate negotiations, surprisingly little attention has been given to the question of what mitigation policies they should follow.

The analysis in this chapter stresses that their own emissions are a relatively small part of the global problem (at least for CO_2) and—as a distinct point, reflecting their low-income levels—that appropriate carbon charges in these countries may well be lower than in advanced economies. At the same time, however (and beyond the element of realpolitik in the expectation of mitigation action to secure climate finance from developed economies), their inclusion in the wider architecture for dealing with climate change is needed both to avoid undoing emission reductions elsewhere and to exploit some of the cheapest options for mitigation. The latter, in particular, could be an opportunity for these countries to mobilize additional resources. Not the least important consequence of considering mitigation policies in these countries, however, is that it places attention on their current systems for taxing energy and fossil fuel use—which, in many cases, can offer them scope for significant gain, even apart from the climate benefit to the wider world.

Appendix: Additional Tables and Charts

Table 6A.1. Country Income Classification

High income				
OECD	Non-OECD	Upper-middle income	Lower-middle income	Low income
Australia	Andorra	Albania	Angola	Afghanistan
Austria	Aruba	Algeria	Armenia	Bangladesh
Belgium	Bahamas	American Samoa	Belize	Benin
Canada	Bahrain	Antigua and Barbuda	Bhutan	Burkina Faso
Czech Republic	Barbados	Argentina	Bolivia	Burundi
Denmark	Bermuda	Azerbaijan	Cameroon	Cambodia
Estonia	Brunei Darussalam	Belarus	Cape Verde	Central African Republic
Finland	Cayman Islands	Bosnia and Herzegovina	Congo (Brazzaville)	Chad
France	Channel Islands	Botswana	Côte d'Ivoire	Comoros
Germany	China, Hong Kong SAR	Brazil	Djibouti	Congo, Dem. Rep.
Luxembourg	China, Macao SAR	Ecuador	India	Kyrgyz Republic
Poland	Croatia	Jamaica	Kosovo	Mali
Greece	Curaçao	Bulgaria	Egypt, Arab Rep.	Eritrea
Hungary	Hungary	Chile	El Salvador	Hungary
Iceland	Equatorial Guinea	China	Fiji	Gambia, The
Ireland	Faeroe Islands	Colombia	Georgia	Guinea
Israel	French Polynesia	Costa Rica	Ghana	Guinea-Bissau
Italy	Gibraltar	Cuba	Guatemala	Haiti
Japan	Greenland	Dominica	Guyana	Kenya
Korea, Rep. of	Guam	Dominican Republic	Honduras	Korea, Dem. Rep.
Netherlands	Isle of Man	Gabon	Indonesia	Liberia
New Zealand	Kuwait	Grenada	Iraq	Madagascar
Norway	Liechtenstein	Iran, Islamic Rep.	Kiribati	Malawi
Portugal	Malta	Jordan	Lao PDR	Mozambique
Slovak Republic	Monaco	Kazakhstan	Lesotho	Myanmar
Slovenia	New Caledonia	Latvia	Marshall Islands	Nepal
Spain	North Mariana Islands	Lebanon	Mauritania	Niger
Sweden	Oman	Libya	Micronesia, Fed. Sts.	Rwanda
Switzerland	Puerto Rico	Lithuania	Moldova	Sierra Leone
United Kingdom	Qatar	Macedonia, FYR	Mongolia	Somalia
United States	San Marino	Malaysia	Morocco	Tajikistan
	Saudi Arabia	Maldives	Nicaragua	Tanzania
	Singapore	Mauritius	Nigeria	Togo
	Sint Maarten (Dutch)	Mayotte	Pakistan	Uganda

High income				
OECD	Non-OECD	Upper-middle income	Lower-middle income	Low income
	St. Martin (French)	Mexico	Papua New Guinea	Zimbabwe
	Trinidad and Tobago	Montenegro	Paraguay	
	Turks and Caicos Islands	Namibia	Philippines	
	United Arab Emirates	Palau	Samoa	
	Virgin Islands (U.S.)	Panama	São Tomé and Príncipe	
		Peru	Senegal	
		Romania	Solomon Islands	
		Russian Federation	Sri Lanka	
		Serbia	Sudan	
		Seychelles	Swaziland	
		South Africa	Syrian Arab Republic	
		St. Kitts and Nevis	Timor-Leste	
		St. Lucia	Tonga	
		St. Vincent and the Grenadines	Turkmenistan	
		Suriname	Tuvalu	
		Thailand	Ukraine	
		Tunisia	Uzbekistan	
		Turkey	Vanuatu	
		Uruguay	Vietnam	
		Venezuela	West Bank and Gaza	
			Yemen, Rep. of	
			Zambia	

Source: World Bank (http://data.worldbank.org/about/country-classifications/country-and-lending-groups).
Note: OECD = Organization for Economic Cooperation and Development.

Table 6A.2. Distribution of Energy Sources, 1980–2008

	1980	1990	2000	2008	Change (Percent) 1980–2008[a]
		Share of energy from petroleum			
World	46.1	39.1	39.2	34.7	–11.4
High income	48.3	42.5	41.9	40.1	–8.3
Upper middle income	41.2	32.6	34.0	27.1	–14.1
Lower middle income	56.1	49.4	38.8	35.3	–20.8
Low income	27.2	26.6	32.9	32.7	5.6
		Share of energy from coal			
World	24.8	25.7	23.2	28.2	3.4
High income	21.3	21.5	18.8	18.3	–3.1
Upper middle income	30.7	31.7	30.7	41.4	10.7
Lower middle income	24.2	26.5	26.8	29.9	5.7
Low income	52.4	49.7	30.1	23.1	–29.3
		Share of energy from natural gas			
World	19.1	22.2	23.3	22.5	3.3
High income	18.6	20.3	22.4	23.8	5.1
Upper middle income	21.6	26.7	25.2	20.5	–1.0
Lower middle income	8.6	14.4	23.8	22.6	14.0
Low income	3.1	7.6	16.5	23.1	20.0
		Share of energy from other sources (nuclear, renewables, etc.)			
World	10.0	12.9	14.3	14.7	4.7
High income	11.7	15.7	16.9	17.9	6.2
Upper middle income	6.5	9.0	10.1	10.9	4.4
Lower middle income	11.1	9.7	10.7	12.1	1.1
Low income	17.4	16.2	20.4	21.1	3.7

Source: Energy Information Administration (http://www.eia.gov/countries/data.cfm) and authors' estimates.
[a] Percentage change for consumption and percentage point change for share

Table 6A.3. Unweighted Average Gasoline Prices, 1998–2010

	1998	2002	2006	2010
All countries	0.40	0.46	0.86	1.23
High income	0.57	0.63	1.04	1.46
Upper middle income	0.30	0.38	0.75	1.12
Lower middle income	0.29	0.34	0.74	1.08
Low income	0.43	0.48	0.91	1.26
Exporters	0.31	0.37	0.66	0.94
High income	0.37	0.44	0.65	0.90
Upper middle income	0.26	0.30	0.53	0.82
Lower middle income	0.25	0.27	0.60	0.91
Low income	0.34	0.45	0.87	1.11
Nonexporters	0.43	0.50	0.95	1.36
High income	0.64	0.69	1.17	1.64
Upper middle income	0.34	0.45	0.89	1.31
Lower middle income	0.31	0.39	0.82	1.19
Low income	0.45	0.49	0.92	1.30
Memorandum items:				
Crude oil prices				
US$ per barrel	13.0	25.0	64.9	79.5
US$ per liter	0.08	0.16	0.41	0.50
Gasoline prices (US$ liter)				
with U.S. markup[a]	0.12	0.23	0.61	0.75
with minimum EU markup	0.24	0.45	1.18	1.44

Source: GIZ (http://www.gtz.de/en/themen/29957.htm) and authors' estimates.

[a] Markups are measured as of 2010 and held constant in percentage terms in other years.

Figure 6A.1. Gasoline Prices in Upper-Middle-Income Countries, December 2010
(U.S. cents per liter, oil exporters in red)

Source: GIZ (http://www.gtz.de/en/themen/29957.htm).

Figure 6A.2. Gasoline Prices in High-Income Countries, December 2010
(U.S. cents per liter, oil exporters in red)

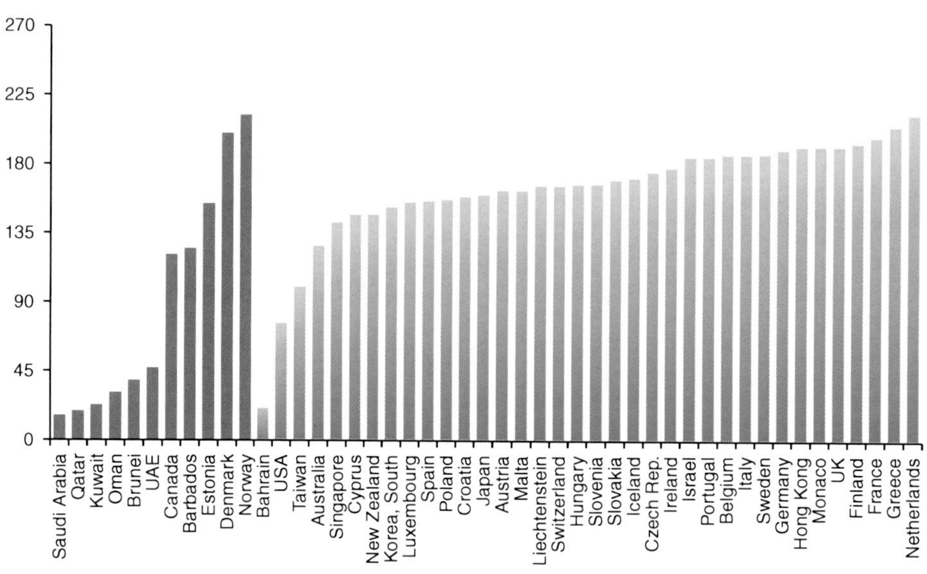

Source: GIZ (http://www.gtz.de/en/themen/29957.htm).

References and Suggested Readings

Useful data on energy prices and use are in:

Deutsche Gesellschaft für Internationale Zusammenarbeit (GIZ), 2012, "International Fuel Prices," available at http://www.gtz.de/en/themen/29957.htm.

Energy Information Administration, 2012, "International Energy Statistics," available at http://www.eia.gov/countries/data.cfm.

U.S. Energy Information Administration, 2008, *A Primer on Gasoline Prices* (Washington: Energy Information Administration). Available at http://www.eia.doe.gov/bookshelf/brochures/gasolinepricesprimer/index.html.

A useful source for emissions projections and assessment of other trends is:

Organisation for Economic Cooperation and Development, 2011, *OECD Environmental Outlook to 2050* (Paris: OECD).

That global efficiency does not require uniform carbon taxes in the absence of international transfers is shown in:

Chichilnisky, G., and G. Heal, 1994, "Who Should Abate Carbon Emissions? An International Perspective," *Economic Letters,* Vol. 44, pp. 443–449.

For experience with fuel subsidies and the development of better-targeted support mechanisms:

Arze del Granado, Javier, David Coady, and Robert Gillingham, 2010, "The Unequal Benefits of Fuel Subsidies: A Review of Evidence for Developing Countries," IMF Working Paper 10/202 (Washington: International Monetary Fund).

Coady, David, Robert Gillingham, Rolando Ossowski, John Piotrowski, Shamsuddin Tareq, and Justin Tyson, 2010, "Petroleum Product Subsidies: Costly, Inequitable, and Rising," IMF Staff Position Note No. 10/05 (Washington: International Monetary Fund).

Grosh, Margaret, Carlo del Ninno, Emil Tesliuc, and Azedine Ouerghi, 2008, *For Protection and Promotion: The Design and Implementation of Effective Safety Nets* (Washington: World Bank).

IMF, 2008a, "Food and Fuel Prices—Recent Developments, Macroeconomic Impact, and Policy Responses" (Washington: International Monetary Fund). Available at www.imf.org/external/np/exr/foodfuel/index.htm.

———, 2008b, "Fuel and Food Price Subsidies: Issues and Reform Options" (Washington: International Monetary Fund). Available at www.imf.org/external/pp/longres.aspx?id=4293.

On how to assess proper levels of fuel taxation, with application, see the following:

Parry, Ian, and Kenneth Small, 2005, "Does Britain or the United States Have the Right Gasoline Tax?" *American Economic Review*, Vol. 95 (September), pp. 1276–1289.

For a framework for thinking about tax design to address local pollution, carbon, and motor vehicle externalities, with rough calculations of the relevant externalities (including in some developing countries):

Parry, Ian, John Norregaard, and Dirk Heine, forthcoming, "Environmental Tax Reform: Principles from Theory and Practice," *Annual Review of Resource Economics*.

There are relatively few studies of carbon pricing design and impact outside more advanced economies. Important exceptions—the first a very early analysis focused on Pakistan and including assessment of the impact on local externalities, the second stressing the importance of labor market distortions—are:

Shah, Anwar, and Bjorn Larsen, 1992, "Carbon Taxes, the Greenhouse Effect, and Developing Countries," Policy Research Working Paper WPS 957 (Washington: World Bank).

Devarajan, Shantayanan, Delfin Go, Sherman Robinson, and Karen Thierfelder, 2009, "Tax Policy to Reduce carbon Emissions in South Africa," Policy Research Working Paper 4933 (Washington: World Bank).

For general issues in VAT design and implementation:

Ebrill, Liam, Michael Keen, Jean-Paul Bodin, and Victoria Summers, 2001, *The Modern VAT* (Washington: International Monetary Fund).

Chapter

7 Fiscal Instruments for Climate Finance

Ruud de Mooij and Michael Keen
Fiscal Affairs Department, International Monetary Fund

Key Messages for Policymakers

- Developed economies have pledged to generate, from 2020, US$100 billion per year to help finance climate mitigation and adaptation in developing economies. This chapter discusses the potential role of fiscal instruments in raising this "climate finance."
- Mobilizing public funds for climate finance does not in principle require earmarking revenue from any particular tax for this purpose, nor does it require using some "innovative" tax. It will also not ensure that funds raised are in addition to other development assistance, or that the finance is found in the most efficient way.
- "Traditional" taxes—such as the value-added tax or personal income tax—could be used to raise additional finance (or expenditure cuts), with much scope for base-broadening in many developed economies, but this would not easily be cast as a common surcharge.
- Most attention has, nonetheless, focused on identifying novel sources of finance linked explicitly to climate finance.
- Prominent among such new sources is the possibility of using some of the revenue raised by comprehensive carbon pricing—which also has a key role to play in catalyzing the private part of climate finance. As a source of earmarked finance, this has the appeal of a particular salience to climate issues. But it is the mitigation benefits of carbon pricing that remain its primary rationale and make it likely to be a particularly efficient source of additional revenue.
- Another promising way to generate revenue, while mitigating emissions, is by removing remaining fossil fuel subsidies in developed economies.

- Charges on fuels used by international aviation and maritime activities are a particularly attractive possible source of finance, given the current absence of any charges or limits on these emissions and the borderless nature of the activities. Finding ways to ensure that such charges do not in themselves adversely affect developing economies, which may be important to obtain sufficiently broad participation to make these changes workable, is feasible.

At the UN Climate Summits of Copenhagen in 2009 and Cancun in 2010, developed countries agreed on a collective commitment to provide resources for climate adaptation and mitigation in developing economies. The sums committed approach US$30 billion for the period 2010–12 ("Fast Start Finance") and rise to US$100 billion per year by 2020.[1] In Cancun, governments also decided to establish the Green Climate Fund to support climate mitigation and adaptation projects in developing economies.[2]

These agreements do not specify, however, where the money is to come from. Addressing this, the UN Secretary General's High Level Advisory Group on Climate Change Financing (AGF) reported in November 2010 on potential sources of revenue for climate finance beyond 2020. The AGF concluded that meeting the goal is challenging but feasible, and that funding will need to come from a wide variety of sources—public and private, bilateral and multilateral—and using a range of instruments. Drawing on and extending that analysis, in October 2011, the World Bank, in collaboration with the IMF and other organizations, reported further on this issue in response to a request from the G-20 Finance Ministers.[3]

Still, it remains an open question as to how governments are to realize their collective commitment. One broad outstanding question is the intended balance between public and private finance. While this will be largely a political decision, it seems clear that their commitments will require governments in developed countries to mobilize at least some new public funds. It is this aspect of climate finance that this chapter addresses: How might the public funds necessary to make a substantial contribution to

[1] The countries committed to providing fast start finance are the 27 EU member states, Australia, Canada, Iceland, Japan, Liechtenstein, New Zealand, Norway, Switzerland, and the United States.

[2] The purpose of the Green Climate Fund (GCF) is to distribute the funds for climate finance, not to raise them; so the GCF is not equivalent to the more general concept of climate finance. At the Durban conference in 2011, it was agreed that the GCF would be an autonomous body within the United Nations, with the World Bank being an interim trustee of the funds over the next 3 years.

[3] The G-20 comprises Argentina, Australia, Brazil, Canada, China, the European Union, France, Germany, India, Indonesia, Japan, Italy, Mexico, the Republic of Korea, Russia, Saudi Arabia, South Africa, Turkey, the United Kingdom, and the United States.

governments' share of the $100 billion commitment be raised? This $100 billion amounts to about 0.25 percent of their joint GDP, so if public resources were to constitute, say, 40 percent of this, the sum required would be approximately 0.1 percent of GDP. While this may not seem huge, it is about half of all current overseas development assistance. And the current severe strains on public finances in many developed economies amplify the evident difficulty of raising such amounts. In the hope of cutting through this difficulty, the reports of the AGF and to the G-20 pay particular attention to "innovative sources of finance"; that is, new tax instruments that could be employed to generate the necessary resources.

In reviewing the continuing debate on raising public funds for climate finance, this chapter relies heavily on background work for the October 2011 report to the G-20 mentioned earlier. It first considers the rationale for climate finance and then discusses the role of fiscal policy in this context. The next two sections discuss traditional domestic revenue sources and some "innovative" sources. Finally, we consider more closely one such possible source that emerges: charges on the use of international aviation and maritime fuels.

Why Climate Finance?

Transfers from developed to developing countries can promote both fairness and efficiency in addressing the collective challenges of climate change.

On ethical grounds, such transfers are particularly salient in the context of climate change. The past emissions that have led to high concentrations of greenhouse gases (GHGs) in the atmosphere have come predominantly from developed economies. Climate finance can be seen as a compensation for the (current and prospective) damages that developed economies have consequently caused in developing economies—which face large needs for costly adaptation in limiting the harm from climate change (perhaps in the order of US$90 billion a year by mid-century according to the World Bank) and will face great remaining residual damages.

Such transfers can also help facilitate the international cooperation that is vital to obtain efficient outcomes. Mitigation policies will be less effective if some countries do not participate in emissions reduction. For example, if carbon were to be priced by only a subset of countries, then emission reductions there would, to some extent, be offset by increased emissions by nonparticipants, whether as a consequence of relocation of emission-intensive activities or in response to a reduction in the world prices of fossil fuels induced by participants' carbon pricing (Sinn, 2012). Emissions reduction in developing and emerging economies—most notably in China and India—is particularly important because a large part of future emissions growth is expected to occur in these countries, and, moreover, many low-cost mitigation options arise in these countries: Failure to exploit these would be a source of significant inefficiency, making the realization of emission reduction much more costly than it need be.

But developing economies, of course, are concerned that shouldering the costs of mitigation and adaptation will hinder their economic growth. By separating who finances climate action from where it occurs, flows of climate finance from developed to developing economies can play a key role in encouraging developing economies to participate in global action. They can, that is, play the pivotal role in reconciling economic efficiency with equity.

The Role of Fiscal Instruments

Fiscal instruments can play two broad roles in mobilizing climate finance: (1) catalyzing private climate finance and (2) raising public funds for transfer to developing economies.

Catalyzing Private Finance

The dominant scale and scope of global private capital markets and the growing fiscal challenges in many developed economies suggest that the financial flows required for a successful climate stabilization effort must, in the long term, be largely private in composition. With properly structured incentives, private initiatives will play an essential role in seeking out and implementing the least-cost options for climate mitigation and adaptation. And the most powerful way to create these incentives, in relation to mitigation activities, is by establishing strong and credible carbon pricing in developed economies—coupled either with similar pricing elsewhere (as, technically, would be preferred) or with international offset provisions that allow covered firms in developed economies to exploit abatement opportunities in developing economies instead of paying carbon taxes or purchasing emissions allowances. This would provide appropriate price signals to leverage the necessary private finance for investment in mitigation and low-carbon investments, including, not least, in developing economies.

Public Finance

By transferring public funds from developed to developing economies, public climate finance raises issues similar to those associated with development aid more generally: (1) How can one ensure that resources are indeed additional to other funds provided and (2) how should the burden be shared across developed economies?

"Additionality" refers to the extent to which new resources add to the existing level of resources flowing from developed to developing economies—in the form of development aid—instead of replacing any of them. Making additionality operational is politically and analytically very difficult, however. One reason is that adaptation needs are often broadly similar to wider developmental ones: Developing more resilient crops, or strengthening social support systems, delivers adaptation benefits, but would have beneficial developmental effects even in the absence of climate change. Difficulties also arise in defining a reference against which "greater" can be determined: Other forms of assistance might not fall, for instance, but simply increase less rapidly than they otherwise would have. Importantly, while a focus on the "innovativeness" as an indicator of additionality is perhaps natural, novelty of a revenue source simply does not resolve this fundamental difficulty.

In mobilizing funds, governments could agree on burden sharing among themselves, whatever revenue source(s) they adopt. A country's contribution could, for instance, be based on its GNP or GDP, reflecting the principle of "capacity to pay," or on population size, reflecting equal rights to the atmosphere. Alternatively—or in combination with other factors—contributions could be based on current or past emissions of GHGs, reflecting "responsibility to pay." With agreement on some such formula, public revenue mobilization could simply come from whatever is each country's most preferred domestic revenue source, separating the provision of some amount of climate finance from the issue of how that revenue is to be raised. By leaving it open to countries how best to raise whatever revenue is asked of them, this might indeed be the most efficient way in which to generate climate finance.

The alternative approach is to agree on the common deployment of some particular tax instrument(s) and earmark the revenue from them, partly or wholly, to climate finance.[4] For instance, a surcharge could be imposed on the value-added tax (VAT) or on personal and corporate tax rates, with the revenue earmarked for climate finance, or part of the revenue from a coordinated carbon tax might be earmarked. Under this approach, burden sharing is determined implicitly—absent further adjustment—by the distribution of the corresponding tax base(s) across countries. Where tax bases differ widely across countries, however—as they commonly do—the idea of a common surcharge becomes much less simple than it may sound. Moreover, rate increases through surcharges might not be the most efficient way to raise additional funds in many countries, as base-broadening measures might be more desirable.

[4] Another possibility, of course, is that developed economies could reduce other expenditures to free up funds for climate finance.

Earmarking of this kind is generally resisted by public finance analysts, the classic objections being that it introduces undesirable inflexibility of spending if it constrains spending patterns or is meaningless (and misleading) if it does not. These objections may, however, have less force in the climate finance context to the extent that there is a clear monetary target to be met. And earmarking may have some appeal in overcoming resistance to new charges, although whether explicitly allocating revenues to the benefit of other countries will be conducive to public support is not obvious. It is, in any case, this earmarking approach that has so far dominated policy debate in this area, and in the rest of this chapter, we focus on possible instruments.

"Traditional" Revenue Sources

What "traditional" revenue sources—meaning ones based on existing tax instruments—would be most appropriate as means for developed economies to generate additional contributions for climate finance? Identifying the best national revenue source requires applying the standard criteria in judging taxes of equity, efficiency, and ease of implementation. All taxes come along with costs of administration and compliance, all affect the income distribution, and distort behavior (in investment, employment, or consumption choices, for instance). While there is general agreement on these principles, there is little consensus on how the tax design best balances between them. Empirical work has, however, led to some broad views on desirable directions of tax reform in advanced economies, which we briefly summarize below. For a much fuller treatment, with some country detail and sense of the revenue potential from measures along these lines, see IMF (2010a).

Value-Added Tax

The VAT has proved to be a relatively efficient source of revenue and is widely believed to generate less costly distortions than many other taxes. Almost all developed economies—except Saudi Arabia and the United States—have a VAT and, on average, it raises for them revenue of over 5 percent of GDP. In many countries, however, exemptions and excessive rate differentiation reduce the effectiveness of the VAT. By changing relative prices of products and services, these measures distort patterns of consumption (and, in some cases, production, too), while achieving few equity gains that could not be better achieved by using the more directly targeted instruments (such as social benefits and earned income tax credits) that are available in advanced economies. Moreover, they also increase costs of administration and compliance. Hence, there is generally substantial scope for further improving the revenue performance of the VAT by reducing exemptions and eliminating reduced rates, combined, if needed, by measures to address any adverse distributional effects. In a number of countries, there is also a sizable "compliance gap"—that is, a loss of revenue due to noncompliance of various kinds, suggesting that improved enforcement would generate substantial revenue gains while also improving the fairness of the tax system. In countries with very low rates, such as Japan, a higher VAT rate could add to revenue. In countries without a VAT, its introduction is a leading option for substantially enhancing revenues.

Corporate Income Tax

The corporate income tax (CIT) is not a likely candidate as a source of additional revenue. International tax competition over the past decades has intensified and has led to significant reductions in statutory CIT rates. Many countries have, at the same time, broadened their tax bases by adjusting tax depreciation rules and restricting deductions (of, for instance, interest expenses). There remains scope for such base-broadening in some countries—by, for example, having leaner tax depreciation allowances or scrapping specific investment allowances—but the potential revenue gains are fairly modest. International coordination could perhaps strengthen the revenue potential of the CIT, but this remains contentious in principle (sensitive issues of tax sovereignty arise, and some see tax competition as a good way to discipline governments in their revenue raising) and quite remote in practice.

Personal Income Tax

The personal income tax (PIT) is generally considered key to achieving equity objectives (because the average rate of tax can be designed to rise with income levels) and might have some potential for higher revenue in a number of developed economies, although this is not likely through rate increases. High effective marginal rates of PIT can have damaging incentive effects on both real activity and compliance. For instance, while incentive effects on labor supply of primary workers are generally found to be modest, tax effects on the participation decisions of secondary workers (mainly married females) can be substantial. Moreover, high tax rates for low-wage earners tend to create large labor market distortions by driving unskilled workers out of the formal labor market. There is also significant evidence that higher rates of PIT encourage tax avoidance and evasion, particularly for high-income individuals. These considerations do not necessarily point to applying low PIT rates across the board, such as envisioned (except perhaps on the lowest incomes) under many "flat tax" proposals, but rather to a progressive structure, combining rates that rise with income and targeted tax credits that address particular incentive or fairness concerns. But in the absence, for instance, of substantially enhanced international cooperation in the taxation of high wealth individuals, whose physical and financial mobility across countries makes it hard for any country to tax them in isolation, scope for raising top marginal tax rates is limited. Where some countries do have scope for raising additional revenue from the PIT, this is done by base-broadening and simplification, such as reducing allowances and exemptions. In the United States, for example, the fiscal cost of tax expenditures under the income tax—a prominent example being mortgage tax relief of a kind that the United Kingdom, for example, has successfully phased out—has been put at over 7.5 percent of GDP[5] (although by no means do all of these tax expenditures not serve a useful purpose).

Property Tax

Recurrent property taxes are a promising source of increased revenue for a number of countries. Efficiency and fairness arguments strongly favor their increased use in many developed economies: They appear to have only limited effects on growth and to be borne mainly by the wealthy. At present, their revenues amount to about 3 percent of GDP in Canada, the United Kingdom, and the United States, but well below 1 percent in other developed economies. This suggests significant untapped potential, but realizing this requires overcoming practical obstacles such as administrative complexities and the unpopularity of these taxes, no doubt partly reflecting their transparency. Nonetheless, property taxation has clear potential for significant and relatively efficient revenue enhancement in several countries.

Summary

There is then scope for raising additional revenue from a variety of traditional sources. But the best way to do this varies across countries and often takes the form of base-broadening rather than simple rate increases. Given too the wide variation in bases, a common add-on to some bases—a surcharge on PIT at the same rate in all countries, for instance—is unlikely to yield a pattern of revenue tailored to broader views on burden sharing or to exploit the most promising options for domestic tax reform.[6] If (as seems implicit in much of the debate) the aim is not only to present to the public with an explicit link between climate finance and some tax instrument feeding it but also to use an instrument that looks the same across developed economies, earmarking additional revenues raised from traditional instruments may not be the best approach.

[5] U.S. National Commission on Fiscal Responsibility and Reform (2010).

[6] Countries could, of course, simply impose such a charge as a signal that a contribution to climate finance is being made, but ultimately contribute some different amount—but this would hardly be transparent.

Innovative Sources of Finance

There are many possible revenue sources that can be called "innovative." Atkinson (2003), for instance, discusses a variety of novel sources of development funding, including global lotteries, the creation of new special drawing rights, charges on remittances, and premium bonds. We restrict ourselves here to three other sources, of varying degrees of novelty: carbon pricing, the removal of fossil fuel subsidies, and financial sector taxes.

Carbon Pricing

Comprehensive carbon pricing policies, such as a carbon tax or emission trading with full auctioning of allowances, are widely viewed as a promising option for climate finance. Chapters 1, 2, and 8 discuss these policies and their optimal design in detail. These chapters, however, view carbon pricing policies primarily as an instrument for efficient climate mitigation and the raising of revenue for general fiscal purposes. Carbon pricing is indeed more effective at reducing emissions than regulatory instruments, providing incentives for clean technology development and promoting international carbon markets. Moreover, carbon pricing is vital for catalyzing private climate finance, as we have discussed.

In the context of climate finance, however, carbon pricing is also motivated specifically as a source of the public revenue needed. A carbon price of US$25 per tonne of carbon dioxide (CO_2) in developed economies, for instance, could raise about US$250 billion in 2020. There is, of course, no logical necessity to earmark funds from carbon pricing for climate finance: The revenues could instead flow into national budgets. It is obvious, but easy to forget, that revenues from carbon pricing can only be spent once. One common concern with carbon pricing, for instance, is the impact on low-income families and/or on the competitiveness of certain industries; prominent among the options for addressing this are lowering other taxes or increasing social benefits to compensate certain groups—but such compensation reduces the net revenue from carbon pricing. Another concern is with the economic distortions caused by the wider tax system. To minimize these, it is generally advantageous to use the revenues from carbon pricing to cut other taxes that distort incentives for work or investment (see Chapter 2), which would again reduce the net revenue that could potentially be used for climate finance. Still, even allocating 10 percent of the sum above for climate finance would meet a quarter of the $100 billion funding commitment.

Removing Fossil Fuel Subsidies

Many countries have subsidies on the production or consumption of fossil fuels. As a revenue source for climate finance, scaling back fossil fuel subsidies in developed economies has attracted particular attention. A recent Organization for Economic Cooperation and Development (OECD) study estimates that fossil fuel subsidies in developed economies amounted to about US$40–$60 billion per year in 2005–10. These subsidies—over half of which were for petroleum, and a little under a quarter of which were for coal and natural gas each (in 2010)—include direct transfers of funds, selective tax reductions or exemptions, and other market interventions that affect cost or prices. At the Pittsburg Summit in 2009, the leaders of the G-20 countries committed to phase out over the medium term their fossil fuel subsidies.

There is thus significant revenue (and environmental improvements) to be found in scaling back these subsidies. As with the traditional instruments, however, making a link between this and the provision of climate finance that is both meaningful and transparent—if that is felt to be important—would not be easy.

Financial Sector Taxes

New taxes on the financial sector have been proposed as a way to raise money for climate finance. "Bank taxes"—typically levied on some subset of banks' liabilities or assets—have been introduced by several countries since the 2008 financial crisis. Their revenue yield, however, is relatively low and seems unlikely to increase substantially (meaning that a large portion of revenues would need to be earmarked to provide a significant contribution to climate finance). In Europe for example, 14 countries have introduced bank levies of some kind, with a revenue yield typically between 0.1 and 0.2 percent of GDP. Other options are discussed in broader public debates. Most prominent are a broad-based financial transactions tax (FTT)—levied on the value of a wide range of financial transactions—and a financial activities tax (FAT)—levied on the sum of the wages and profits of financial institutions. Both were considered and compared extensively in the IMF's 2010 report to the G-20 on financial sector taxation. Broadly speaking, the FTT has acquired greater political momentum (notably with a formal proposal by the European Commission, with an estimated revenue of 0.5 percent of GDP), while the FAT has acquired greater support from tax policy specialists. For instance, expert opinion tends to be that an FTT would increase the cost of capital and so could have a significant adverse impact on long-term economic growth and that its real burden would likely fall on final consumers rather than on actors in the financial sector. FTTs are also particularly vulnerable to avoidance and evasion. The FAT, on the other hand, is in large part simply intended to help correct distortions caused by the exemption of financial services under the VAT. Both the FTT and the FAT, nonetheless, are technically feasible—with the appropriate degree of international cooperation—and both could raise significant revenues. As a revenue source, however, they are not necessarily global: Revenue could instead flow into national budgets, and burden sharing for climate finance could be based on other factors as well.

International Transportation

Without the more comprehensive carbon pricing along lines discussed above, pricing the emissions from international aviation and maritime fuels—either through carbon taxation or emissions trading schemes with allowance auctions—has been proposed as an innovative source of climate finance. These two sectors account for about 1.5 and 2–3 percent, respectively, of global CO_2 emissions, and some projections have their combined share rising to 10–15 percent by 2050, if unchecked. But these emissions are excluded from explicit charges and are not included in the Kyoto Protocol—reflecting precisely that the very nature of these activities makes it unclear which country should charge or regulate the fuel they use. And that, in turn, makes it natural to think of imposing charges on them and turning the revenue so raised to some collective use.

The Case for Charging Emissions in These Sectors

International aviation and maritime activities are currently taxed relatively lightly from an environmental perspective: Unlike domestic transportation fuels, they are subject to no excise taxes that can reflect environmental damages in fuel prices. Moreover, these sectors receive favorable treatment from the broader fiscal system. For instance, shipping income generally receives favorable tax treatment, being subject to relatively low "tonnage" taxes rather than normal corporate taxation. And international passenger flights are, unlike the generality of consumption items, almost invariably exempt from VAT.

Pricing emissions is widely viewed as the most economically efficient and environmentally effective instrument for tackling climate challenges in these sectors. Under the auspices of the International Maritime Organization (IMO) and the International Civil Aviation Organization (ICAO), both sectors are taking important steps to improve both the fuel economy of new planes and vessels and the efficiency of routes and speeds. In the maritime industry, an agreement was reached in July 2011 within the IMO on the first mandatory GHG reduction regimen for an international industry. However, higher fuel prices resulting from pricing instruments would be still more effective. Beyond reinforcing these efforts, they would, for example, reduce the demand for transportation (relative to trend) and promote retirement of older, more polluting vehicles.

The principles of efficient pricing design are the same in these as in other sectors. For emissions taxation, this means minimizing exemptions and targeting environmental charges on fuels rather than imperfect proxies, such as passenger tickets or arrivals and departures. For emissions trading, it means auctioning allowances to provide a valuable source of public revenue, incorporating provisions to limit price volatility, and developing institutions to facilitate trading markets.

A globally implemented carbon charge of US$25 per tonne of CO_2 on fuel used could raise about $12 billion from international aviation and about $25 billion from international maritime transport annually in 2020. Revenues would be higher if, in addition to addressing environmental considerations, charges were also set to address other aspects of undertaxation noted above. A $25 tax is expected to reduce CO_2 emissions from each sector by roughly 5 percent, mainly by reducing fuel demand. Compensating developing economies for the economic harm they might suffer from such charges—ensuring, in a phrase widely used in this context, that they bear "no net incidence"—is widely recognized as critical to their acceptability, and how this might be done is discussed below. Such compensation seems unlikely to require more than, say, 40 percent of global revenues, which would leave about $22 billion or more for climate finance or other uses.

Failure to price emissions from one of these sectors should not preclude pricing efforts for the other. Though commonly discussed in combination, the two sectors are not only different in important and fiscally relevant respects (e.g., ships mainly carry freight, while airlines primarily serve passengers), they also compete directly only to a limited degree. Nonetheless, simultaneous application to both is clearly preferable and could enable a common charging regimen (enhancing efficiency) along with, perhaps, a single compensation scheme for developing economies.

International Cooperation

Extensive cooperation would be needed in designing and implementing international transportation fuel charges—especially for maritime transport—to avoid revenue erosion and competitive distortions. Underlying the current tax-exempt status of international transportation fuels are fears that unilateral taxation would variously harm local tourism and commerce, undermine the competitiveness of national carriers, raise import prices and/or reduce the demand for exports. Fuelling might take place in countries without similar policy measures, so that even the revenue gain might be compromised. To overcome these fears, some degree of international coordination is likely to be needed.

In the case of international aviation, even an agreement with substantially less than universal coverage—for example, one that exempted some vulnerable developing countries—could still have a significant effect on global emissions and revenue potential, given the relatively limited possibilities for carriers to simply refuel wherever taxes are lowest. For maritime bunker fuels, globally comprehensive pricing is more critical: Large vessels can much more easily avoid a charge by taking up fuel in any countries where such charges do not apply.

Incidence and Compensation

The importance of ensuring "no net incidence" of these charges for developing economies has been stressed by many. Achieving this requires careful consideration of what precisely is the "real" incidence of these charges—that is, of who it is that suffers a consequent loss of real income. This can be quite different from who bears legal responsibility for the payment of the charge; in these sectors, these two groups may very well even be resident in different countries. It is the real incidence that matters for potential compensation, and this is sensitive to views on demand-and-supply responses; in the present context, it will also vary across countries according to their shares of trade by sea and air, the importance of tourism, and so on.

The first step in determining the incidence of these charges is assessing their impact on fuel prices. Jet and maritime fuel prices might not rise by the full amount of any new charge on their use, as some portion of the real burden is likely to be passed back to refiners and/or producers of oil products. If, however, it is fairly easy for refiners to shift production from jet and maritime fuels to other petroleum products (as may be plausible, given possibilities for reconfiguring refineries over the longer term), then the amount refiners have to absorb will be relatively small.

Even with full pass-through to fuel prices, however, the impact on final prices of aviation services and landed import prices of goods carried by ship—and on the profitability of the aviation and maritime operators—is unlikely to be large. A charge of US$25 per tonne of CO_2 might raise average air ticket prices by about 2 to 4 percent (and then, of course, the total cost of a tourist package by a still smaller proportion) and the price of most seaborne imports by about 0.2 to 0.3 percent. The fairly modest scale of these effects means that the real burden on both the ultimate users and the providers of international aviation and shipping services is likely to be small—and the latter, in any case, reflects a scaling back of unusually favorable fuel tax treatment for these industries rather than the introduction of unfavorable treatment.

Nonetheless, there may be a need to provide adequate assurance of no net incidence on developing countries by providing explicit compensation.

Significant challenges arise in designing such a scheme because of the jurisdictional disconnect between the points at which a charge is levied and the resulting economic impacts—especially for maritime transport. Practicable compensation schemes require some verifiable proxy for the economic impact as a key for compensation. Fuel take-up may provide a good initial basis in aviation, and simple measures of trade values may have a role in relation to maritime.

Fully rebating aviation fuel charges for developing economies (or giving them free allowance allocations) would be a promising way to protect them from the adverse effects of fuel charges. Indeed, this could more than compensate them: that is, they might be made better off by participating in such an international regime (even prior to receiving any climate finance). This is because much of the real incidence of charges paid on jet fuel disbursed in developing economies (especially tourist destinations, the impact on which has emerged as a particular concern) would likely be borne by passengers from other (wealthier) countries.

In contrast, there can be less confidence that rebating charges on maritime fuel taken up in developing economies would adequately compensate most of them. Unlike airlines, shipping companies cannot be expected to normally tank up when they reach their destination. Some countries—hub ports like Singapore—disperse a disproportionately large amount of maritime fuel relative to their imports, while the converse applies in importing countries that supply little or no bunker fuel, including landlocked countries. Revenues from charges on international maritime fuels could instead be passed to or retained in developing economies in proportions that reflect their share in global trade. Although this is relatively straightforward to administer, further analysis is needed to validate whether this approach would provide adequate compensation, for example, for countries that import goods with relatively low value per tonnage.

Implementation

Implementing globally coordinated charges on international aviation and/or maritime fuels would raise significant governance issues. New frameworks would be needed to determine how and when charges (or emissions levels) are set and changed, to provide appropriate verification of tax paid or permits held, to govern the use of funds raised, and to monitor and implement any compensation arrangements. While the European Union experience on tax coordination indicates that agreements can be reached, it also shows the sensitivity of the sovereignty issues at stake in tax setting and collection. One possibility is to limit the need for a separate decision process by linking an emissions charge on international transportation to the average carbon price of the largest economy-wide emission reduction scheme. There may be some role for the ICAO and IMO, with their technical expertise in these sectors, in implementing these charges.

The familiarity of operators and national authorities with fuel excises suggests that implementation costs would be lower with a tax-based approach than with an emissions trading scheme. Collecting fuel taxes is a staple of almost all tax administrations and very familiar to business; implementing trading schemes is not. Ideally, taxes would be levied to minimize the number of points to control—which usually means upstream in the production process. If taxation at the refinery level is not possible, the tax could be collected where fuel is disbursed from depots at airports and ports or directly from aircraft and ship operators.

Policies might be administered nationally, through international coordination, or in some combination of the two—with the appropriate institutions for monitoring and verification depending on the approach taken. For example, national governments might be responsible for implementing fuel charges or trading schemes on companies distributing fuel to airlines or ships. An alternative for the maritime sector might be to collect the charge without the intervention of national authorities along lines similar to the present International Oil Pollution Compensation funds administered by the IMO.

For international aviation, legal issues loom large in considering possible charges: The current fuel tax exemptions are built into multilateral agreements within the ICAO framework and bilateral air service agreements, which operate on a basis of reciprocity. Amending the Chicago Convention and associated resolutions would remove these obstacles, although the EU experience on intra-union charging seems to suggest the possibility of overcoming them without doing so. An alternative approach would be to use an emissions trading system or scheme (ETS) in this sector, and indeed, the EU is including international aviation in the EU ETS beginning January 1, 2012. This, however, is proving contentious and is currently the subject of litigation. No such legal difficulties appear to apply, however, in relation to international maritime fuels.

Conclusion

Climate-related transfers to developing economies can serve the objectives of both fairness and, by helping to realize low-cost mitigation opportunities there, efficiency. Developed economies could mobilize the additional public revenues needed by a range of measures, including through traditional domestic tax instruments, although differences in the bases of these taxes mean that a common surcharge is not likely to be acceptable or appropriate.

"Innovative" financial instruments are not necessary to mobilize climate finance; but, nonetheless, the search for such instruments has been central to the debate.

Among these, its salience to climate issues has led to suggestions that some of the revenue from comprehensive carbon pricing in developed economies (which, given appropriate offset and other institutional structures, can also stimulate private financial flows to developing economies) be allocated to this end. This, of course, would detract from the appeal that such pricing may have in addressing deep fiscal challenges in many of these countries. Charges on fossil fuels used in international maritime and aviation industries, for which a strong environmental case can be made, have some advantage in the climate finance context—especially perhaps the former—with regard to the difficulty of allocating the base to particular countries and the need for widespread cooperation if they are to be effective. Measures can likely be found to protect developing economies from the adverse effects of such charges (which, in most cases, seem likely to be modest), but significant legal issues remain in establishing effective fuel charges for international aviation.

References and Suggested Readings

Advisory Group on Climate Change Financing, 2010, *Report of the Secretary-General's High-Level Advisory Group on Climate Change Financing*, November 5, 2010 (New York: United Nations).

Atkinson, A. B., 2003, "Innovative Sources for Development Finance," in *New Sources of Development Finance*, ed. by T. Atkinson (Oxford: Oxford University Press).

IMF, 2010a, "From stimulus to consolidation: revenue and expenditure policies in advanced and emerging economies," Policy Paper, April (Washington: International Monetary Fund).

———, 2010b, "A Fair and Substantial Contribution by the Financial Sector," presented at the G-20 Toronto Summit, Toronto, Canada (June 26–27).

———, 2011, "Promising Domestic Fiscal Instruments for Climate Finance" (Washington: International Monetary Fund). Available at http://www.imf.org/external/np/g20/.

——— and World Bank, 2011, "Market-Based Instruments for International Aviation and Shipping as a Source of Climate Finance" (Washington: International Monetary Fund and World Bank). Available at http://www.imf.org/external/np/g20/.

Sinn, H., 2012, *The Green Paradox: A Supply-Side Approach to Global Warming* (Cambridge, Massachusetts: MIT Press).

U.S. National Commission on Fiscal Responsibility and Reform, 2010, "The Moment of Truth." Available at http://www.fiscalcommission.gov.

World Bank, 2011, "Cost to Developing Countries of Adapting to Climate Change: New Methods and Estimates," Consultation draft (Washington).

World Bank, International Monetary Fund, Organisation for Economic Cooperation and Development, African Development Bank, Asian Development Bank, European Bank for Reconstruction and Development, European Investment Bank, and Inter-American Development Bank, 2011, "Mobilizing Climate Finance," Paper prepared at the request of G-20 Finance Ministers, October 6. Available at http://climatechange.worldbank.org/content/mobilizing-climate-finance.

CHAPTER

8 Carbon Pricing: Lessons Derived from Experience*

Tom Tietenberg
Colby College, United States

Key Messages for Policymakers

- Although replacing regulatory emissions control policies with market-based instruments has produced significant cost savings, the predominant effect has been to reduce emissions. The savings and emission reductions have fallen somewhat short of their full potential, however, partly because actual designs have deviated from the most economically efficient designs (e.g., because programs are not fully comprehensive). Market-based policies have also promoted clean technology investments (although gains are not always as large as expected). Carbon leakage effects to date have been relatively modest.
- Emissions pricing programs often take the form of "hybrid" schemes that combine upstream with downstream systems and emissions taxes with emissions trading. For example, in Australia and the European Union, large downstream emitters are covered by cap-and-trade systems (which are a more natural extension of other earlier environmental regulations), while more diffuse sources (e.g., home heating fuels, transportation fuels) are covered by taxes. These hybrid systems can still cover most energy-related carbon dioxide (CO_2) emissions and can be reasonably cost-effective, at least if there are not big differences in emissions prices across sectors.
- Although the Kyoto Protocol sought to simultaneously control six greenhouse gases (GHGs) by translating them into a common index of CO_2 equivalents, no existing program covers all these gases. For administrative ease, most programs focus solely on energy-related CO_2 emissions, although this may not be a major drawback given that CO_2 accounts for about three-quarters of global GHGs. Moreover, many programs are now beginning to transition to a more comprehensive coverage of gases.
- Price volatility has been a bigger concern than market power in trading systems to date (though experience is limited to developed economies). Cap-and-trade systems

*This chapter was originally a survey paper prepared for the International Monetary Fund.

often contain price volatility through provisions for permit banking (allowing entities to save permits for later use when expected allowance prices are higher) and advance auctions (allowing entities to buy allowances at current prices for use in several years). Permit borrowing (which allows entities to use permits before their designated date) is more restricted (due to a fear that firms might default on owed allowances), but this does not seem to have been a problem.

- Revenues from carbon taxes and auctioned allowances have been used for reducing other taxes, compensating industries, offsetting regressive impacts on households, and promoting renewable and energy efficiency programs. Use of revenues for industry compensation has diminished over time, however, with greater appreciation of the value of forgone revenues and tendency to overcompensate (in fact, power producers reaped windfall profits in the early phases of the EU trading scheme). Some programs (e.g., Australia) address adverse effects on low-income households with progressive adjustments to the broader tax system.
- Emissions "offset" provisions are a common means for reducing the financial burden of carbon pricing programs on sources. While introducing an offset program can result in larger total emissions reductions under a carbon tax, it will not affect total emissions reductions under a cap-and-trade. But the challenge is to ensure that the credited emissions reductions outside of the formal program can be measured and would not have occurred anyway (without the offset credit). Due to concerns about credibility, most programs impose limits on offsets, but newer approaches attempt to distinguish between more credible offsets (which are allowed) and less credible ones (that are rejected).
- Price and emissions transparency are important for accountability, reducing the likelihood of fraud and facilitating programmatic refinement over time. Independent evaluations are an important component of ex post evaluation of pricing programs, and data access is needed for outside reviewers to perform these evaluations.

Although programs to control climate change based on pricing carbon are relatively new, programs to price pollution more generally are not. Various forms of emissions trading and pollution taxes or charges have been around since at least the late 1960s (Table 8.1).

Both types of programs provide a wealth of experience from which to draw insights on how well these programs work and how the context matters. The experience with these programs also sheds considerable light on the consequences of design choices, given the vast array of available options. This brief survey is designed to summarize some of the chief lessons that can be drawn from this experience.

Table 8.1. Selected Existing Air Pollution Fee or Emissions Trading Systems

Emissions Trading	Air Emissions Fees
Traditional Pollutants	***Traditional Air Pollutants***
The U.S. Lead Phase-Out Program (Lead, 1985)	Japan (Sulfur Oxides, 1968)
The U.S. Sulfur Allowance Program (SO_2,1990)	China (Multiple Pollutants, 1982)
Santiago, Chile (Particulates, 1992)	France (Multiple Pollutants, 1985)
California Regional Clean Air Incentives Market	Sweden (Nitrogen Oxides, 1992)
(NO_x and SO_x, 1994)	Taiwan Province of China (Multiple Pollutants, 1996)
Eastern U.S. NO_x Budget Program (NO_x 2003)	
Climate Change	***Climate Change***
The Kyoto Protocol's Clean Development and	Finland (1990)
Joint Implementation Mechanisms (2005)	Netherlands (1990)
The European Union Emissions Trading Scheme (2005)	Sweden (1991)
U.S. Northeastern States Regional Greenhouse Gas	Norway (1991)
Initiative (2009)	United Kingdom's Climate Change Levy (2001)
New Zealand (2010)	Denmark (2005)
	Switzerland (2007)
	Quebec, Canada (2007)
	British Columbia, Canada (2008)
	India (2010)

Source: Author.

This chapter opens with a short summary of five programs that run the gamut of program types. These include two carbon tax programs (British Columbia, Canada, and Sweden), two cap-and-trade programs (Europe and the northeastern United States), and a hybrid that includes both (Australia). In the next section, we isolate some of the lessons for system design that emerge from actual experience, followed by a section in which we identify the lessons about how well these programs perform in practice, using such metrics as cost savings, emissions reductions, market transformation, and technology innovation and diffusion. We then provide an in-depth summary of the lessons for policymakers.

Providing Context: A Brief Look at Five Illustrative Carbon Pricing Programs

Swedish Carbon Tax Program

In Sweden, carbon is taxed both directly on each emitted unit of CO_2 and indirectly (and imperfectly) via an energy tax on fossil fuels that is not based on their carbon content. The carbon tax was introduced in Sweden in 1991 as a complement to the existing system of energy taxes, and the existing taxes were simultaneously reduced by 50 percent.

When the European Union Emissions Trading Scheme (EU ETS) was introduced in 2005, some sectors were covered both by the carbon tax and by the EU ETS. Most emissions from the transport sector and from households were excluded from the EU ETS, but were covered by other taxes. To avoid double regulation, the government decided to exempt the industries covered by the EU ETS from the carbon tax. Hence, all sectors are now covered by a carbon price, but it varies greatly across firms and sectors, with some activities being fully exempted. Although it is covered by the EU ETS, electricity production faces neither energy nor carbon taxes, but a special electricity consumption tax is levied on households.

This mixed system has been effective in reducing emissions. According to the Swedish Ministry of the Environment, Swedish greenhouse gas (GHG) emissions fell by almost 17 percent during the period from 1990–2009, mainly through fuel substitution. The share of renewable energy has increased from 34 percent in 1991 to 44 percent in 2007 and is among the highest in Organization for Economic Cooperation and Development (OECD) countries. Electricity is now almost CO_2 free with hydroelectricity and nuclear power accounting for more than 90 percent of electricity generation.

The carbon tax is considered to have caused emission reductions mainly in the residential sector (largely by promoting district heating, which is more efficient than localized heating) and has diminished the historic trend of increasing emissions in transport. Experts believe that the carbon tax's impact on industry is probably small due to the many exemptions granted.

British Columbia Carbon Tax Program

Implemented in 2008, the carbon tax in British Columbia, Canada, is imposed on each tonne of CO_2 equivalent (CO_2-e) emissions from the combustion of each fuel. In this case, CO_2-e is the amount of CO_2, methane, and nitrous oxide (N_2O) released into the atmosphere. The non-CO_2 emission levels are adjusted to a CO_2 equivalent basis using global warming factors. Fuel for commercial aviation as well as for cargo and cruise ships is exempted. This program covers an estimated 77 percent of total GHG emissions in British Columbia.

Administratively, the carbon tax is applied and collected at the wholesale level in essentially the same way that the province collects its motor fuel taxes, a strategy that makes administration easier. The tax is ultimately passed forward to consumers.

All revenue generated by this revenue-neutral carbon tax is returned to Canadians residing in British Columbia through tax cuts. The first two personal income tax bracket rates were reduced by 5 percent on

January 1, 2008. To attempt to protect low-income households, the Low Income Climate Action Tax Credit program provides adult residents with lump sum tax credits that are reduced by 2 percent of net family income over specified income thresholds.

European Union Emissions Trading Scheme

Launched in 2005, the EU ETS is the largest emissions trading system in the world. The EU ETS now operates in 30 countries (the 27 EU member states plus Iceland, Liechtenstein, and Norway). It covers CO_2 emissions from installations such as power stations, combustion plants, oil refineries, and iron and steel works, as well as factories making cement, glass, lime, bricks, ceramics, pulp, paper, and board. Between them, the installations currently in the scheme account for almost half of the EU's CO_2 emissions and 40 percent of its total GHGs.

The EU ETS established a cap on the total amount of certain GHGs that can be emitted by liable entities. Within this cap, companies receive emission allowances, which they can sell to or buy from one another as needed. The number of allowances is reduced over time so that total emissions fall. By 2020, emissions are targeted to be 21 percent lower than those of 2005.

While the scheme currently covers only CO_2 emissions, the scope of the ETS will soon be extended to include other sectors and other GHGs such as N_2O and perfluorocarbons. Pending the outcome from legal challenges, airlines are expected to join the scheme in 2012. Expansion to the petrochemicals, ammonia, and aluminum industries and to additional gases is expected in 2013. In 2013, the scheme will also begin to auction off allowances, with the ultimate goal being to attain full auctioning by 2027.

Regional Greenhouse Gas Initiative

In 2009, ten states—Connecticut, Delaware, Maine, Maryland, Massachusetts, New Hampshire, New Jersey, New York, Rhode Island, and Vermont—launched the first market-based regulatory program to reduce GHG emissions in the United States. Through the Regional Greenhouse Gas Initiative (RGGI), each participating state caps CO_2 emissions from power plants, auctions CO_2 emission allowances subject to a price floor, and invests the proceeds in strategic energy programs that may further reduce emissions.

In retrospect, this program imposed a rather weak cap, but expectations about tighter caps in the future, coupled with very favorable natural gas prices that made a rapid transition to lower carbon fuels cost-effective, apparently led to dramatic declines in emissions. By the end of 2010, the second year of the program, emissions were 27.1 percent lower than the cap and 25.6 percent lower than the year the program was announced (2005). While the recession played some role in this decline, economic analysis suggests that the main source of the reductions was fuel substitution, promoted by lower natural gas prices. During the period 2005–10, electricity generation from residual fuel oil fell 95 percent, generation from coal decreased 30 percent, and natural gas generation increased by 35 percent.

RGGI states allocate a large proportion of the money specifically to promote energy efficiency (e.g., weatherizing buildings and incentivizing investments in new technologies offering much more energy-efficient lighting). Analysis has found that energy efficiency is now the most cost-effective tool for reducing GHG emissions in this region in the near term.

The politics of using the money for energy efficiency is illustrated by noting what happened to public support for the program in the states within the RGGI that (in direct contradiction to program intensions) diverted money that was originally targeted at energy efficiency to the general treasury for deficit reduction. New York took $90 million during the fall of 2010, roughly half of its fund; New Jersey zeroed out its fund, taking all $65 million; and New Hampshire, a much smaller state, took $3.1 million. Since diverting the funds undermined a major source of political support for these programs, diversion triggered a withdrawal from RGGI by New Jersey. An attempt to withdraw by the New Hampshire legislature was only thwarted when the governor vetoed the legislation.

RGGI has a price floor (approximately $2 per tonne), which is indexed to remove the effects of inflation, and (due to the surplus of allowances created by the large emissions reductions relative to the cap) that floor has been binding. In fact, the floor has played a major role in the region because revenue from the auctions is an important source of funding for the popular energy efficiency programs. Without that floor, a considerable amount of funding would have been lost, thereby creating a source of instability in the regional energy efficiency market.

Australian Hybrid System

The Australian plan envisions a two-stage transition from a carbon tax to an emissions trading market:

- In the first stage, lasting from July 1, 2012, until June 30, 2015, emitters will face a fixed price for each tonne of CO_2 emitted. The price will start at $A23 (US$23.81) per tonne and will rise at 2.5 percent per annum in real terms.

- On July 1, 2015, the fixed carbon price regime will transition to a fully flexible price regime with the price determined by the emissions trading market.

While half of a source's compliance obligation must be met through the use of domestic permits or credits, after the commencement of the flexible price period, offsets from credible international carbon markets and emissions trading schemes may be used to fulfill the remaining obligation.

A price ceiling and floor will apply for the first 3 years of the flexible price period. The price ceiling will be set at $A20 (US$20.70) above the expected international price and will rise by 5 percent in real terms each year. The price floor will be $A15 (US$15.53), rising annually by 4 percent in real terms.

While coverage under the emissions trading market will be quite broad, it will not be universal. Four of the GHGs covered by the Kyoto Protocol—CO_2, methane, NO_x, and perfluorocarbons (in this case from aluminum smelting)—will be covered. Transport fuels will be excluded, but an equivalent carbon price will be applied to domestic aviation, domestic shipping, rail transport, and nontransport use of fuels. No carbon price will apply to household transport fuels, light vehicle business transport, or off-road fuel use by the agriculture, forestry, and fishing industries. Agricultural emissions will also be exempted.

Of the potential revenue derived from 2011–15, more than 50 percent will be used to reduce the cost burden on households. Assistance will be delivered through the tax and transfer system. To reduce the regressivity of the burden, the government will also target the assistance to low- and middle-income individuals by more than tripling the statutory tax-free threshold for personal income taxes.

Some revenue has also been specifically targeted to facilitate the transition for highly impacted firms. An Energy Security Fund will oversee two main initiatives: (1) payments for the closure of around 2,000 megawatts of coal-fired generation capacity by 2020 and (2) transitional assistance to highly emissions-intensive, coal-fired power stations in the form of a limited free allocation of Australian carbon permits and cash until 2016/17, estimated to be worth $A5.5 billion.

Other initiatives that use revenue generated by the program will focus on job promotion and promoting both renewable energy and energy efficiency.

Lessons about Program Design

Countries in the process of designing carbon pricing systems have several design options to consider. In many of these cases, the choices are similar regardless of whether the preferred mechanism is a carbon tax or emissions trading, but in others, the choices are unique to each instrument.

Chapter 2 summarizes the design issues from an optimality perspective. In this chapter, we summarize the actual design choices made by existing programs and the lessons that can be derived from their experience.

Instrument Choice

Older public discourse frequently framed the instrument choice as deciding whether carbon taxes were superior to emissions trading or vice versa. In practice, it is now more common to frame the issue in terms of how they can best be combined.

For example, these two instruments can be used sequentially, where one instrument is used initially until a transition to the second instrument is completed. Some evidence, which we will discuss, suggests that emissions trading markets take a while to mature. To make sure that these early markets do not produce volatile or unstable prices, it is possible to start with a tax regime that produces known, stable prices until such time as participants become sufficiently familiar with abatement choices and their costs that an emissions market can take over. Australia's plan to impose a carbon tax to take effect in mid-2012, for example, is expected to pave the way for a transition to a carbon trading system where prices would be determined by the market by mid-2015.

These instruments can also effectively be used at the same time but applied to different sectors. It is not a coincidence that emissions trading systems, such as the EU ETS and RGGI, target larger sources, while taxes have been typically targeted on more diffuse sources such as residential or transport emissions. In the Australian emissions trading plan, for example, sources not covered by the cap will be controlled by an equivalent carbon price. Similarly in Sweden, household and transport emissions are controlled largely via taxation and large enterprise emissions are controlled via the EU ETS. This can be perfectly consistent with optimality as long as the resulting prices in the two systems are, in fact, equivalent

Finally, the two instruments can be combined to form a hybrid instrument. One well-known example, the use of a price collar with emissions trading, is mentioned in Chapter 1, and we also discuss it more in the section in this chapter on price volatility.

Scope of Coverage: Gases and Sources

Gases. The Kyoto Protocol envisions that all six named GHGs could be simultaneously controlled with a carbon pricing program by translating all six of them into a CO_2-e.[1] In practice, this is accomplished using their relative Global Warming Potentials (GWPs), which are defined as the cumulative radiative forcing effects of a unit mass of gas relative to CO_2 over a specified time horizon (commonly 100 years). In an inclusive system, it is this CO_2-e that forms the basis for either the tradable commodity or the tax base.

Including multiple gases can reduce the cost of reaching a specific concentration target by quite a bit. The disadvantages include not only that the GWP approach does not produce a perfect equivalency (residency time in the atmosphere varies considerably among the different gases, for example), but also the fact that some of these gases may be more difficult to monitor.

In practice, for administrative ease, most programs currently focus on CO_2 emissions from fossil fuel consumption. This can be a viable transition strategy since it is not difficult to add additional gases as the monitoring capacity matures. Both RGGI and the Swedish Carbon Tax, for example, have a CO_2-only focus. However one jurisdiction, the Bay Area Air Quality Management District in San Francisco, California, does tax CO_2-e (all gases), and Australia plans to cover four of the six Kyoto GHGs.

Historically, some regulatory targets other than CO_2 have also been adopted. For example, Boulder, Colorado, taxes only electricity use. A number of European countries also tax energy (Btus) and/or electricity consumption.[2]

While a uniform tax on energy does promote energy conservation, it fails to promote switching among fuels with different emissions per unit of energy.

Sources. From the point of view of minimizing cost, more universal coverage of sources (like gases) is also better. Yet no existing program involves universal control. As noted above, the EU ETS covers only certain categories of large emitters, although expansion to other sectors is in process. RGGI covers only one sector (large power generators).

---

[1] The six primary GHGs are CO_2, methane, NO_x, hydrofluorocarbons, perfluorocarbons, and sulfur hexafluoride.

[2] Btus are British thermal units. This metric is used to provide a comparable energy measure across different fuel types.

Aside from including some sectors and not others, source size commonly affects coverage. For most programs, even in covered sectors, only facilities over a specified size threshold usually incur pricing obligations.

Expanding the original scope of coverage has occurred as a result of broadening the concept of where the carbon pricing can be applied. In general terms, emissions can not only be controlled directly (via "downstream" targeting), but also indirectly (via "upstream" targeting), or even via a hybrid involving some combination of the two.

A downstream point of regulation would focus control on the point of use, where GHGs are emitted into the atmosphere. An upstream system imposes the taxes or allowance requirements on the point of extraction, production, import, processing, or distribution of substances, which, when used or combusted, would generate GHGs.

The downstream approach is probably the most common in practice. For example, in the Bay Area Air Quality Management District, the fees are applied directly to the emitting facilities, and in RGGI, it is the emitting generators (and not the fuel suppliers) that are required to submit allowances. In Australia, the carbon pricing mechanism will apply directly to several hundred of their biggest polluters.

Like so many other design choices, the point of regulation should not be considered simply as a binary choice. Hybrids involving upstream control of some sources and downstream control of others have proved to be popular options and are becoming more common. For example, the British Columbia carbon tax is generally applied and collected upstream, except for the fee on natural gas, which is collected at the retail level. Overlaps (double taxation) are commonly avoided either by rebates or granting within-category exemptions to those facing the possibility of double taxation.

Temporal Flexibility

Emissions trading systems offer more temporal flexibility by allowing banking, borrowing, and advance auctions. *Banking* means holding an allowance for use beyond its designated year. *Borrowing* means using an allowance before its designated date. *Advance auctions* sell allowances that can be used after some future date, commonly 6 or 7 years hence.

The economic case for allowing this temporal flexibility is based upon the additional options it allows sources in timing their abatement investments, which lower compliance costs. The optimal time to install new abatement equipment or to change the production process can vary widely across firms. Factors such as the age of the equipment that is being replaced and the number of available technological options for additional control clearly matter.

Price considerations also argue for temporal flexibility. Forcing firms to adopt new technologies at exactly the same time concentrates demand at a single point in time (as opposed to spreading it out). Concentrated demand would raise prices for the equipment as well as for the other complementary resources (such as skilled labor) necessary for its installation.

Banking also has been shown to reduce the damage caused by price volatility. Storing permits for unanticipated outcomes (such as an unexpectedly high production level, which triggers higher than expected emissions) can reduce the future uncertainty considerably. Because stored permits can be used to achieve compliance during tight times, they provide a safety margin against unexpected contingencies.

The existence of a banking system, where allowances can be stored for future use, may also contribute to the political durability of the policy. Enacting a well-designed carbon pricing policy will not be sufficient if it cannot be maintained when public attention wanes and the policy faces the threat of being undermined or distorted by special interest politics. Even just the credible threat of dismantlement can have a strong negative effect on investment incentives.

History suggests that reforms are sustainable when the major participants have an interest in their continuation and, in general, when policy preservation incentives are aligned in a way that is self-reinforcing. Those holding banked allowances (as well as entities involved in and profiting from the infrastructure of the carbon market, such as brokerage houses, registries, etc.) are likely to insist on preserving a stable market.

In recognition of these substantial advantages, banking is widely used in emissions trading programs. Borrowing, on the other hand, has experienced more limited use, in part due to a fear that firms that borrow heavily could become enforcement problems later. The Australian ETS plans to allow limited borrowing of permits such that, in any particular compliance year, a covered source can surrender permits from the following vintage year to discharge up to 5 percent of its obligation. So far, the fact that borrowing has been limited has not seemed to raise costs in any significant way.

Advance auctions are now more common, but little analysis of their impact has made it into the literature.

Using Revenue from Taxes or Auctions

Carbon taxes and auctioned allowances not only provide incentives for reducing emissions, but they raise revenue as well. The distribution of the revenue from auctioned allowances or carbon taxes can, in principle, enhance policy efficiency, reduce the regressivity of the distribution of the financial burden, and/or improve the political feasibility and stability of the program, but those benefits depend upon what is done with the revenue. Operating programs provide experience with a wide variety of choices, which can be helpful in seeing whether experience matches expectations.

Containing the Burden on Target Groups. It is not uncommon for nations setting up carbon pricing programs to have to deal with powerful political concerns about their possible economic impacts, especially on energy-intensive businesses. These concerns have resulted in several design strategies that use potential or actual revenue to contain possible cost increases faced by these businesses, including exemptions, preferential tax rates, or gifted allowances.

Exemptions are a common strategy for targeted burden containment (especially in European tax systems). Types of exemptions include (1) exempting all emissions from sources that emit fewer emissions than some established threshold (a strategy followed by most programs), (2) exempting emissions from sources that are covered by another policy to prevent double taxation (also common), (3) exempting emissions from sources deemed unacceptably vulnerable to cost increases, and (4) exempting emissions where international legal issues introduce special implementation barriers. While the first two types of listed exemptions may not normally raise significant cost-effectiveness issues, exemptions of the third and fourth types can. Because facilities that receive exemptions face no controls on GHGs from that instrument, their incentive to reduce emissions is eliminated. Furthermore, when the instruments are designed to reach specific quantitative targets, the other facilities must pick up the slack created by exemptions, which raises compliance costs for the program as a whole.

Preferential tax rates are even more common. In Norway, for example, the pulp and paper industry, fishmeal industry, domestic aviation, and domestic shipping of goods pay reduced rates. In Sweden, manufacturing, agriculture, cogeneration plants, forestry, and aquaculture face lower rates.

Another possibility (with mixed results) for reducing the cost burden on vulnerable firms, issuing rebates, is illustrated by the Swedish Nitrogen Charge System (Box 8.1).

Gifting can occur in either a tax system or an ETS. With a tax system, it involves taxing only emissions above the gifted threshold, a strategy found in some European effluent charge systems. Alternatively, in emissions trading, some proportion of the allowances in a cap-and-trade system can be gifted (given free of charge) to favored sectors. Either approach eliminates the financial burden associated with paying for gifted emissions, but in contrast

Box 8.1. Rebates in Action: Mixed Consequences Flowing from the Swedish Nitrogen Charge System

The Swedish Nitrogen Charge took a different approach to cost containment. It was intended from the beginning to have a significant incentive effect, not to raise revenue. Although the charge rate is high by international standards (thereby producing an effective economic incentive), the revenue from this tax is not retained by the government, but rather is rebated to the emitting sources (thereby reducing the impact of the tax on competitiveness).

It is the form of this rebate that makes this an interesting scheme. While the tax is collected on the basis of emissions, it is rebated on the basis of energy production. In effect, this system rewards plants that emit little NOx per unit of energy and penalizes plants that emit more NOx per unit of energy. This approach provides incentives to reduce emissions per unit of energy produced, but not to reduce the amount of energy. Hence, it reduces fewer total emissions than an unrebated tax.

Over the period 1992–2005, the average emission intensity was nearly cut in half, but total output of useful energy from participating plants increased by more than 70 percent (due to expanding energy demand). As a result, total NOx emissions from the units targeted by the nitrogen charge barely fell.

Source: Sterner, and Turnheim (2008).

to an exemption, it does not relieve the sector of its obligation to control GHGs.

The EU ETS and California, at least initially, gifted some or all of the allowances to parties based upon some specified eligibility criteria. In the European Union, free allowances are allocated based on product-specific benchmarks for each relevant product. The starting point for the benchmarks is the average of the 10 percent most-efficient installations, in terms of GHGs, in a sector. In California, utilities will apparently receive 90 percent of 2008 electricity sector emissions for free in preparation for the start of the GHG trading scheme on January 1, 2013.

From an efficiency point of view, basing the amount of gifted allocation simply on historical emissions is an inferior basis for allocation since it can end up rewarding sources that have the poorest historical track record (notice how the EU ETS system circumvents this problem). Furthermore, if this method is known in advance, it can even discourage early reduction actions, lest such reductions lower the emitter's subsequent gifted allocation.

The experience in the EU ETS has enriched our understanding of the dynamics of gifted systems. Empirical evidence has demonstrated that in deregulated electricity markets—mainly Germany, the Nordic countries, the United Kingdom, and the Netherlands—a significant share of the value of the gifted allowances in the marketplace was passed through to consumers in the form of higher prices. Since the allowances were gifted, the benefiting firms earned "windfall profits."

Generally, this experience with gifted allocations suggests that they are sometimes necessary for political reasons, but as we will describe, marginal increases in gifting also reduce or eliminate some very beneficial other possibilities for using that revenue. Furthermore, the evidence suggests that the revenue necessary to fully protect sectors that are truly vulnerable is a minor fraction of the total that would be derived from a revenue-raising approach, since most of the burden of carbon pricing is ultimately passed forward to households in higher energy prices.

In recognition of their large and rising opportunity cost, gifting of allowances is being used less as experience with carbon pricing grows. Even in systems granting gifted allowances, the proportion of gifted allowances is usually diminished over time. For example, the EU ETS aims to auction off 20 percent of all EU allowances in 2013, with subsequent gradual increases aiming at auctioning off 70 percent by 2020. The ultimate goal is to attain full auctioning by 2027. In RGGI, gifting already plays a very small role. Approximately 86 percent of CO_2 allowances are offered at auction, and roughly only 4 percent of CO_2 allowances are offered for sale at a fixed price.

Using Revenue to Lower Other Taxes. Considerable variability can be found in the revenue-use choices made by different countries as they have sought to enhance policy efficiency and to reduce inequity. The energy and carbon taxation schemes in several EU member states have been guided by the environmental tax reform. This reform of national tax systems seeks to shift the tax burden from conventional sources, such as labor and capital, to alternative sources such as environmental pollution or natural resource use. As discussed more fully in Chapter 2, using revenue to reduce taxes on items that distort the broader economy considerably reduces the overall costs of the policy. At the same time, this revenue recycling can reduce, at least to some degree (depending on the choice of which distortionary taxes to reduce), the regressivity of the distributional burden of the costs.

The burden of a carbon pricing program is estimated to be regressive, particularly for nontransport emissions in industrialized counties, in the absence of any redistribution of the revenues, because lower-income households use a larger proportion of their earnings to purchase energy-intensive products (gas and electricity being the most important). Gifted allowances intensify this regressivity because the gains accrue to stockholders, who on balance tend to have higher incomes than wage earners.

Countries have made rather different choices in how they have chosen to recycle revenue. Sweden and Finland have mainly recycled revenue by lowering income taxes. On the other hand, Denmark and the United Kingdom have predominantly used revenues to lower employers' social security contributions. British Columbia, Canada, has principally used the revenue to lower personal, corporate, and small business income taxes.

Recognizing that some efficiency-enhancing strategies, such as lowering corporate taxes, do little to diminish regressiveness, some programs target some proportion of the funds specifically to reduce the cost burdens on households, particularly low-income households. In the Australian plan, more than 50 percent of the revenue will be used to reduce the cost burden on households. Cash assistance will be delivered through the tax and transfer system. Assistance will be targeted to low- and middle-income individuals by more than tripling the statutory tax-free threshold. The tax cuts, increases in pensions, and cash transfers have been designed to at least offset any expected average price impact from the carbon price on low-income households. Middle-income households will be eligible for assistance that is expected to meet their average price impact.

Promoting Renewable Energy and Energy Efficiency. Other programs use the revenue to promote renewable energy and energy efficiency. Under a carbon tax, this strategy reduces emissions further, while under an ETS, it may instead lower allowance prices depending on whether these emissions reductions are covered by the cap. In Denmark, although about 60 percent is returned to industry, some 40 percent of tax revenue is used for environmental subsidies. Quebec, Canada, deposits its carbon tax revenue into a "green fund," which supports measures offering "the largest projected reduction in, or avoidance of, GHGs" (Sumner, Bird, and Dobos, 2011, p. 934).

As we have noted, RGGI tends to concentrate its revenue on promoting energy efficiency. These investments not only tend to have a higher cost-effectiveness than renewable resource investments in those states, but by lowering demand, they have even lowered electricity prices (thereby reducing the regressive impact of the policy). These incentivized investments in energy efficiency have also raised the competitiveness of several large industrial facilities and have increased the political support for (and stability of) these programs in the process.

Wise use of the revenue from auctioned allowances or carbon taxes has, in practice, enhanced policy efficiency, reduced the regressivity of the distribution of the financial burden, and/or improved the political feasibility and sustainability of the program. In this case (at least at the macro level), reality is increasingly matching expectations.

Achieving multiple objectives implies making multiple choices on how to use the revenue. Fortunately, the revenue stream seems large enough to allow governments to make headway on all these fronts.

General Cost Containment: The Role for Offsets

Exemptions, differential prices, gifting, and rebates all attempt to reduce the burden for certain targeted sources. Allowing offsets is a common means for reducing the cost on all participants by expanding the supply of reduction possibilities to sources not otherwise covered by the carbon pricing program.

Offsets allow emissions reductions for sources not covered by the cap or not included in the base of a GHG tax to be credited against the cap or tax base by the acquiring party. Offsets or offset tax credits perform four roles in pricing GHGs: (1) by increasing the number of reduction opportunities, they lower the cost of compliance;[3] (2) lowering the compliance cost increases the likelihood of enacting the program; (3) they extend the reach of the program by providing economic incentives for reducing sources that are not covered by the tax or cap;[4] and (4) because offset credits separate the source of financing of the reduction from the source that provides the reduction, it secures some reductions (in developing countries or low-income projects, for example) that for affordability reasons might otherwise be precluded (Box 8.2).

While offset credits can be a permanent component of a carbon pricing program, they may also be used as a transition strategy. For example, as long as countries remain outside the cap, offset credits may represent the best opportunity to secure emission reductions in those countries. Once all countries fall under a pricing regimen, this specific form of offset would become unnecessary.

The challenge for an effective offset program is to ensure that all three of the primary requirements (the reductions should be not only quantifiable, but

[3] The cost effects can be dramatic. Official preliminary estimates from the U.S. Environmental Protection Agency have suggested that the liberal offset provisions in the American Clean Energy and Security Act of 2009 (Waxman-Markey), which passed the House of Representatives but did not become law, would have had the effect of reducing the allowance price by approximately 50 percent.

[4] Current examples from RGGI include credits for reducing methane from landfills or for the additional carbon absorption resulting from reforestation investments.

Box 8.2. Charismatic Carbon Offsets: Extending the Benefits from Carbon Reductions

When producing the carbon reduction in an offset results in an additional social benefit that has substantial public appeal, offsets can be an important source of financing for desirable projects in addition to the carbon reduction that might otherwise not be funded. Because these credits belong to a class of offsets known as "charismatic offsets," they are expected to command a premium in the voluntary market. Consider the two following examples of these charismatic offsets:

In Maine, some 80 percent of the housing stock is heated by oil, and many low-income families cannot afford either to pay the high cost of heating these homes or the cost of weatherizing these homes to make them more energy efficient. Government assistance is available, but it is insufficient to fill the need.

MaineHousing, an independent state agency set up to assist Maine's low- and moderate-income people, embarked a few years ago on an innovative program to use offsets to supply more financing for weatherization projects. When it weatherizes low-income homes, MaineHousing creates carbon savings that will be quantified and certified by the Verified Carbon Standard (note that the RGGI cap that covers Maine deals with electricity, not fuel oil; hence, these savings are outside the cap.) Once the amount of saved carbon is certified, the certified offsets are sold in the voluntary market.

In July, 2011 it was announced that revenue from the certified credits accumulated so far will be sold to General Motors Chevrolet Division with the revenue from this sale being poured back into weatherizing more low-income homes.

Another application of this concept is being developed by organizations such as the World Wildlife Fund. Under the United Nations' Reducing Emissions from Deforestation and Forest Degradation program, forest preservation can result in offsets based upon the certified carbon absorbed by the preserved or reforested trees. While preserving the specific forest underlying these offsets also preserves habitat for charismatic species (tigers, rhinos, etc.), this class of offsets can, in principle, command a price premium in the offset market (based upon the public relations benefits) with the additional revenue used to enhance that habitat. Both the magnitude and sustainability of these charismatic offset price premiums remain to be seen.

Sources: http://www.mainehousing.org/ABOUT/ABOUTGreen/Carbon; http://www.mainehousing.org/news/news-details?PageCMD=NewsByID&NewsID=502; Eric Dinerstein, 2011, "The Future of Conservation," lecture presented at Colby College, Maine, Fall.

also enforceable and additional) are met. One barrier involves the tension over the trade-off between transactions costs and offset validity—ensuring valid offsets is not cheap. Other criticisms involve not only the types of projects being certified (an alleged overemphasis on non-CO_2 gases) and the skewed regional distribution of clean development mechanism (CDM) activity (with Brazil, China, India, and the Republic of Korea creating more than 60 percent of generated credits), but also the amount of subsidy being granted (with incremental costs of reduction of these non-CO_2 gases being well under the price received for a credit) and the adverse incentives created for host countries to pursue reduction on their own (developing counties may well hesitate to undertake projects on their own as long as they can get someone else to pay for them through an offset mechanism such as the CDM).

Motivated by concerns over the validity of offsets, most programs now consider ways to limit their use. One historical method has been to restrict the use of offsets (either domestic or foreign or both) to some stipulated percentage of the total required allowances. In RGGI, for example, CO_2 offset allowances may be used to satisfy only 3.3 percent of a source's total compliance obligation during a control period, although this may be expanded to 5 percent and ultimately 10 percent if certain CO_2 allowance price thresholds are reached. In 2011, Germany announced that it would not allow any offsets to be used to pursue its reduction goals. Similarly, California is seen by observers as being highly unlikely to allow CDM-certified emissions reductions to be used in its emissions trading scheme when it kicks off in 2013. California's position on these CDM offsets contrasts sharply with that of Australia, which apparently plans to rely heavily on the purchase of offsets to hold costs down. Until 2020, Australian sources can use offsets to meet up to 50 percent of their annual liability.

The disadvantage of this quantitative limit approach is that it not only raises compliance costs, but it also fails to distinguish between high-quality and low-quality offsets. Both are treated with the same broad brush.

Newer approaches make these kinds of quality distinctions. Countries could, for example, establish eligibility criteria to identify certain offset types that it deems as acceptable (and therefore treated as fungible with allowances), while not allowing others where the reductions are more speculative and/or the monitoring is less reliable. For example, in 2011, Australia announced that it would not accept HFC 23, or N_2O, offsets from the CDM program.

An alternative approach, being discussed especially with respect to forestry offsets, is to discount the offset (for example, giving certified credit for only 50 percent of the expected reduction) to provide a margin of safety against the uncertainty in the magnitude of the ultimate reductions from this offset project. Discounting can specifically address such concerns as permanence, additionality, and leakage.

Price Volatility

A tax system fixes prices, and in the absence of any administrative intervention to change those prices, price volatility is not an issue. That is not the case with cap-and-trade systems either in principle or practice.

Experience validates the concern that emissions trading can be plagued by volatile prices. The EU ETS, the Regional Clean Air Incentives Market (RECLAIM), and the U.S. Sulfur Allowance Program have all experienced events where prices became quite volatile.

- In the case of the EU ETS, it was attributable to two correctable design mistakes—inadequate public knowledge of actual emissions relative to the cap and a failure to allow allowances in the first phase to be banked for use in the second phase.
- In the case of RECLAIM, it was due to an unanticipated rise in the demand for allowances resulting from an unexpected shortage of important low- or nonpolluting electricity-generating sources (natural gas and hydro from out of state) at precisely at the same time that the program was reaching the "crossover" point, where actual emissions would be expected to exceed allocations unless emission reduction controls were installed at facilities.
- In the U.S. Sulfur-Emissions Trading Program, prices became volatile in the 2004–05 and 2008–09 periods. In the first period, a large rise in allowance prices was triggered by a rapid rise in natural gas prices due in part to Hurricane Katrina, while in the second period, volatility was introduced by two U.S. Circuit Court rulings dealing with a related program (the Clean Air Interstate Rule) to control sulfur.

One potentially appealing, but as yet untested in practice, approach to lower the degree of possible price volatility couples a "price collar," consisting of a safety valve price ceiling and an allowance reserve, with a price floor. Establishing a safety valve ceiling would allow sources to purchase additional allowances at a predetermined price set sufficiently high to make it unlikely to have any effect unless allowance prices exhibited unexpected spikes.[5] To prevent these purchases from breaking the cap, they would come from an allowance reserve that was established from allowances set aside for this

[5] While this program is not designed as a means to generally hold prices down, units covered by the program could seek a design (a low ceiling price) to fulfill that purpose. Without going into the technical details, in general, offsets perform that role much better than price ceilings. Notice, for example (as we have discussed in this chapter) how the RGGI design makes the allowable amount of offsets contingent on allowance prices.

purpose from earlier years, an expansion in the availability of domestic or international offsets, or perhaps from allowances borrowed from future allocations. An allowance reserve has been included in the California program set to begin in 2012 and was part of the Waxman-Markey bill that passed the U.S. House of Representatives but failed to become law.

Australia has proposed a price floor that would apply for the first 3 years of the flexible price period (as we previously discussed). The price floor has two roles: (1) By recognizing that low carbon prices diminish or eliminate incentives to invest in new low carbon forms of energy, it tries to assure investors that prices will fall no lower than the floor, and (2) it provides some price protection for the revenue stream from auctioned allowances. As described above, RGGI also has a price floor, and it has, in fact, constrained prices.

Australia's new climate policy also introduces another innovation targeted at reducing price volatility that can arise in immature markets. During the fixed-price period, sources covered by the program will be able to purchase allowances from the government at a fixed price. Any allowances purchased at the fixed price must be surrendered and cannot be traded or banked for future use. Subsequently, prices would be determined by supply and demand as the carbon pricing system transitions from the fixed-price regime to an emissions trading market.

Policy Evolution via Adaptive Management

One of the initial fears about market-based systems was that they would be excessively rigid (resistant to change), particularly in the light of the need to provide adequate security to investors. Policy rigidity was seen as possibly preventing the system from responding to better information.

Rigidity was certainly not the result with the process set in motion to control ozone-depleting gases under the Montreal Protocol. Not long after the original caps were set, better scientific information confirmed that they were insufficient to achieve the desired results. Soon after this discovery, the cap was made more stringent; the existence of specific caps did not preclude an ability to adjust their stringency to changing conditions in that setting.

Yet system modification, if not done carefully, does have the potential to undermine the incentives that provide the foundation for the system. The stable and predictable prices established by a tax system provide a form of market security investors depend upon when making long-lived energy investments. Similarly cap-and-trade systems depend on allowance holders having secure ownership rights to the allowances. When a new understanding of climate science makes adjustments in the tax rates or the caps necessary, that security can be jeopardized. Can the desire to constantly incorporate change be made compatible with the desire to preserve sufficient security for investors?

While careful design cannot eliminate the tension between flexibility and security, it can reduce it by implementing an adaptive management system coupled with flexible policy instruments. Adaptive management goes beyond trial and error in the sense that it designs initial programs at least in part to learn about the magnitudes and effects of responses so that knowledge from those experiments can be used to improve subsequent programs. This is especially important in a scientifically complex system (such as climate change) that will involve many combinations of policies with equally complex patterns of interaction.

Adaptive management involves two possible pathways—active and passive.[6] Passive adaptive management involves the establishment in advance of triggering thresholds and prescribed rules for changing key parameters once any specified threshold is exceeded, while active adaptive management specifies in advance the deliberative process by which the need for change will be identified and the necessary modifications implemented.

The virtue of *passive adaptive management* strategies is that they do not depend on future political action; the prespecified action follows directly from the prespecified triggering event. The disadvantage is that not all circumstances that affect the optimal choice may be known in advance. Passive strategies are typically less tailored, but they are also less susceptible to political manipulation.

One of the simplest passive strategies, and one that has already been adopted in some tax programs, is to index the tax rate or the level of a price floor or ceiling to a specified rate of inflation index. In this case, measured inflation is the triggering event, and the resulting action is to increase the prevailing price or tax rate to account for the amount of measured inflation. Without this strategy, the real tax rate or price would become weaker and weaker over time. Conversely, under an ETS, the quantity of allowances might be reduced at a fixed annual rate over time.

Experience also provides a guide for how change should *not* be implemented. In one such case in the early days of the first U.S. program, called at the time the Emissions Trading Program, it became clear that additional emissions reductions were needed. Some states simply confiscated all banked credits.

[6] Readers with a sense of economic history will recognize the kinship of this discussion with earlier discussions (monetary policy, for example) involving debates between rules versus discretion.

While this had virtue in that it provided quick, easy reductions, it proved disastrous in the long term because it destroyed any future incentive to create those credits. Since banked credits are a significant source of temporal flexibility for those required to comply with the law, this turned out to be not only short-sighted, but also quite cost-ineffective.

The takeaway lesson from this specific experience is clear—authorized levels of emissions for banked allowances and offsets in general should not be affected by a change in the cap. Government should *not* confiscate banked credits or offsets just because they are handy and not currently in use.

Active adaptive management involves specifying a specific and transparent process for dealing with the evolution of the system over time. The advantage of this pathway is that it can take into consideration all of the circumstances that prevail at the time the change would be needed, while the disadvantage is that discretion allows politics to distort or prevent effective change.

Implementing an active adaptation strategy would involve specifying trigger points (possibly just future dates, but potentially contingencies such as falling below specified horizons for meeting deadlines) for initiating any investigation of the need for modification. Reviews of the appropriateness of the previously specified cap or tax schedule should be conducted on a periodic, announced schedule. The process and criteria for deciding whether the caps or tax rates need to be modified (and, if so, how large a change might be needed) should be made transparent to interested parties in advance.

Some small steps toward this approach can be found in the plan announced by the Commonwealth of Australia. This plan will announce the first 5 years of caps by 2014. For the sixth and subsequent years, the pollution cap will be extended by 1 year every year to maintain 5 years of known caps at any given time. The government has set out in advance the considerations that will be used to set future caps, including such aspects as progress in meeting emissions targets and the social and economic impacts of various choices. A specific independent body, the Climate Change Authority, will be established by legislation to advise on key aspects of the carbon pricing mechanism, not only with respect to the caps, but also on such aspects as the chosen values for the price ceiling and price floors (as we have discussed). The level of the international price will be examined closer to the point of transition to a flexible price to ensure that the price ceiling still reflects the specified margin above its expected level.

The details of climate change policy will necessarily change over time. Markets can accommodate those changes providing the changes are neither arbitrary nor capricious. Establishing a transparent, adaptive management plan can go a long way to facilitating the necessary evolution without unreasonably jeopardizing the incentives upon which the system depends.

Transparency and Accountability

One lesson that emerges quite forcefully from the operational experience with pricing mechanisms is the importance of price and emissions transparency. Not only can this reduce the likelihood of fraud, but also it can enhance market effectiveness. One of the successful features of the U.S. Sulfur Allowance Program—the zero revenue auction—had the effect of reducing the uncertainty associated with trading and facilitated negotiations about price and quantity by making prices public (see Box 8.3). Furthermore, the availability of both organized exchanges (where buyers and sellers could meet) and knowledgeable brokers lowered the transactions costs for those seeking trades.

Box 8.3. Price Revelation: The Sulfur Allowance Program's Zero Revenue Auction

Prior to the sulfur allowance program, price information had generally been private, known only to those engaging in specific over-the-counter trades and their brokers. Lack of information on prices has made abatement decisions much more uncertain and, hence, more difficult. To reduce this uncertainty, the U.S. Sulfur Allowance Program initiated a rather unique auction that would produce public prices, but no revenue.

To supply the auctions with allowances, the U.S. Environmental Protection Agency (EPA) withheld approximately 2.8 percent of the total annual allowances allocated to all units. Private allowance holders (such as utilities or brokers) could also offer their allowances for sale at the EPA auctions.

In these auctions, allowances are sold at the bid price (not a single market clearing price) starting with the highest priced bid and continuing until all allowances have been sold or the number of bids is exhausted. The allowances withheld by EPA are sold before allowances offered by private holders. Offered allowances are sold in ascending order, starting with the allowances for which private holders have set the lowest minimum price requirements.

For our purposes, the key aspect is that the EPA returns all proceeds and unsold allowances on a pro rata basis to those units from which EPA originally withheld them.

Analysis of the prices before and after the auction suggests that the auction was effective in allowing a single market price to emerge and henceforth to promote trading activity.

Source: http://www.epa.gov/airmarkt/trading/factsheet.html and http://www.epa.gov/airmarkt/progress/ARPCAIR10.html

Other forms of transparency are also important. Transparency promotes accountability and facilitates programmatic refinement over time. Independent evaluations are an important component of ex post evaluation, and therefore it is important to provide data access so outside objective reviewers can perform those evaluations. To facilitate these evaluations, it is necessary to provide access not only to the details of the program, but also to various performance measures such as emissions levels for all covered sources.

Most of the programs covered by this review do not meet this standard of transparency. Many of the websites (at least the versions in English) about the programs contain little more than a series of press releases. An early partial exception was the U.S. Sulfur Allowance Program, which provided sufficient access to data from two major external ex post reviews of that program. The RGGI website also lists all the covered sources, their annual individual emissions for a number of years both before and after implementation, and the results of each auction for each participating state and for the program as a whole. The site also contains not only the documents that describe the details of the program, but also evaluations by a consulting firm hired specifically to do a public accounting of whether any evidence of market power has emerged.

The revised ETS Directive adopted in 2009 provides for the centralization of the EU ETS operations into a single EU registry. This Union Registry will be operated by the commission and will replace all EU ETS registries currently hosted in the member states in 2012.

The widespread use of the Internet has made the current possibilities for sharing information historically unprecedented. It is now possible to make very large datasets available to the public at very low cost. The existence of these technological possibilities means that ex post evaluations can no longer be prevented simply because information dissemination is too expensive. Resolving the tension between providing public access to the data and adequately protecting proprietary information is the largest remaining challenge.

Lessons about Program Effectiveness

As we have seen, actual design choices do not always mirror the choices suggested by an optimality perspective. Given that divergence, it is legitimate to ask how well these programs have worked in practice.

Cost Savings

Two types of studies have conventionally been used to assess cost savings and air quality impacts—ex ante analyses based on computer simulations, and ex post analyses, which examine the actual implementation experience.

A substantial majority, though not all, of the large number of ex ante studies of programs involving pollutants other than carbon have found traditional regulatory limits on emissions to be a significantly more expensive way to reduce emissions than the least-cost allocation of the control responsibility. These studies demonstrate that a change from more traditional regulatory measures to more cost-effective market-based measures such as emissions trading or pollution taxes could potentially either achieve similar reductions at a much lower cost or could achieve much larger reductions at a similar cost to more traditional policies based upon source-specific limits.

The evidence also suggests that these two instruments typically produce more emissions reduction per unit expenditure than do other types of existing policies such as renewable resource or biofuel subsidies.

The one exception to this robust finding of significant cost savings arises when the amount of additional reduction is so stringent that the control authority has no choice but to reduce emissions close to the limits established by technological feasibility. In that case, the *immediate* potential cost savings from moving to market-based approaches are typically very small; however, over time, as new technologies are encouraged and introduced by the market-based approaches, those savings can rise considerably.

Although the number of detailed, completed ex post studies is small, the few existing studies typically find that cost savings from moving to these market-based measures are considerable, but less than would have been achieved if the final outcome were fully cost-effective. In other words, while both taxes and emissions trading are fully cost-effective in principle, in practice, they fall somewhat short of that ideal in part because actual designs, fashioned in the crucible of politics, deviate from the dictates of optimality.

Although political manipulation can distort both tax and emissions trading outcomes, emissions trading has a unique potential exposure to price manipulation arising from the presence of market power. Should it materialize, market manipulation could reduce the cost savings.

Two rather distinct types of market power are possible in emissions trading. The first arises when a price-setting source or a collusive coalition of sources seeks to manipulate the price of allowances in order to reduce their financial burden from pollution control. The second stems from the desire of one predatory source or a collusive coalition of sources to leverage market power in either the allowance market or the output market (or both) for increasing profits across both markets. Generally, market power concerns arise only in situations where participants or coalitions of participants control a significant proportion of the market.

Implementation experience with the use of emissions trading has uncovered only one example of market power and that flowed directly from a design flaw.[7] The paucity of cases involving market power is perhaps not surprising since these concerns diminish with large markets and most carbon markets have a large number of participants. However, if emissions trading expands to settings where the market is fragmented and hence limited to a relatively few participants, this could change.

Emission Reductions

While the evidence in general is that implementing these market-based programs reduces emissions (sometimes substantially), that evidence is on less solid ground than the evidence on costs. Almost all of the emissions evidence is based upon what happened to emissions following the introduction of the program relative to what they had been prior to the program (as opposed to comparing them to a counterfactual baseline defined in terms of what would have happened otherwise).

Using emissions patterns before and after implementation is problematic in at least two important senses. First, a historic baseline can be a highly inaccurate benchmark. Suppose, for example, in the absence of the program, the emissions level would have increased dramatically over time. In that case, a program that stabilizes emissions would inappropriately be judged to have accomplished nothing, since emissions would not have declined relative to the year of introduction. In fact, however, it would have achieved substantial reduction relative to what would have happened otherwise. Second, with the exception of the U.S. Sulfur Allowance Program and the U.S. Lead Phase-Out Program (where the evidence is compelling), the degree to which credit for these reductions can be attributed to the market-based mechanisms (as opposed to exogenous factors or complementary policies) is limited.[8]

With that cautionary note in mind, in general, emissions in the noncarbon pollution pricing programs have fallen substantially following the introduction

[7] Evidence uncovered in the RECLAIM market in California found that some generators manipulated NOx emission permit prices during the latter half of 2000 and early 2001. Since higher NOx prices could be used to cost-justify higher bids into the California electricity market, intentionally inflating bids in the RECLAIM market ultimately resulted in higher prices for the electricity produced by the price manipulators. (The higher cost in one market was leveraged into much higher revenues in a related market.) One aspect of the RECLAIM market design that facilitated its use to raise wholesale electricity prices was the "paid-as bid" nature of transactions (as opposed to everyone paying the same market clearing price as happens in RGGI). This design allowed suppliers interested in raising RECLAIM prices on individual transactions to do so without impacting the prices paid by other credit buyers wanting to keep their purchase prices down.

[8] Note, as described above, that the large RGGI reductions, for example, would not have occurred without a simultaneous fall in natural gas prices.

of these market-based mechanisms. (For example, emissions from covered sources have fallen 67 percent in the sulfur allowance program, and lead emissions from gasoline were eliminated.)

Emission reductions following the time of introduction are also the norm for carbon pricing programs, although the reductions in general tend to be more modest. The reductions are typically in the high single digits for the carbon tax, with one country, Norway, actually reporting an increase. Collectively, the EU ETS reported "reduced annual emission per covered installation" by 8 percent from 2005 to 2010 (Hedegaard, 2011).

Not all sources of GHGs end up being regulated, of course, and that raises the specter of leakage. Leakage can occur when pressure on the regulated source to reduce emissions from one location results in a diversion of emissions to unregulated, or less regulated, sources. Common channels for this diversion involves firms moving their polluting factories to countries with lower environmental standards or consumers increasing their reliance on imported products from countries with unregulated sources. For example, states falling under RGGI could potentially import (presumably cheaper) electricity from neighboring states that are not covered by RGGI. In these cases where some emissions are diverted to unregulated areas, the *net* emissions reduction effects of the program (reductions from the regulated sources minus the offsetting increases from the less regulated sources) could be smaller than the more apparent gross effects.

Generally, to date, the evidence suggests that carbon leakage effects have been rather small (typically less than 10 percent).

Market Transformation

Although hard evidence on the point is scarce, a substantial amount of anecdotal evidence suggests that pricing pollution can change the way environmental risk is treated within polluting firms. In the absence of pollution pricing, environmental management was commonly relegated to the tail end of the decision-making process. Specifically, the environmental risk manager was not involved in the most fundamental decisions about product design, production processes, selection of inputs, and so on. Rather, she or he was simply confronted with the decisions already made and told to take whatever precautions were necessary to ensure compliance. This particular organizational assignment of responsibilities inhibited the exploitation of one potentially important avenue of risk reduction—pollution prevention.

Carbon pricing tends not only to move emissions reduction objectives earlier in the decision process, but it also tends to get corporate financial officers involved in environmental risk management. Furthermore, as the costs of compliance rise in general, environmental costs trigger more scrutiny and consideration. Over time, reducing environmental risk can become an important component of the bottom line. Given its anecdotal nature, the evidence on the extent of organizational changes that might be initiated by pricing pollution should be treated more as a hypothesis to be tested than a firm result, but its potential importance is large.

The existing experience also provides some evidence on the behavior of markets. While economic theory treats markets as if they emerge spontaneously and universally as needed, in practice, in unfamiliar markets such as these, the participants frequently require some experience with the program before they fully understand (and behave effectively) in the market. Both regulators and environmental managers of emission sources have apparently experienced considerable "learning by doing" effects in response to pricing pollutants with the result that markets tend to operate much more smoothly after they have been in existence for some time.

Technological Innovation and Diffusion

The literature contains some empirical support for the theoretical expectation that the implementation of market-based mechanisms would induce both emission-reducing innovation and the adoption of new emission-reducing technologies. While the gains in innovation from emissions trading programs have not always been as large as expected, most studies do find a statistically significant response. Furthermore, with respect to the international transfer of mitigation technologies, some evidence also suggests that the CDM, one component of emissions trading, has been a channel for hastening transboundary diffusion, although some observers question the quantitative importance of this channel in practice.

Some case studies also suggest that environmental taxes have made a difference in introducing innovative strategies.

- Norway's carbon tax, for example, has also apparently promoted carbon sequestration, one form of technological innovation.
- The Swedish Nitrogen Charge apparently promoted both innovation (improvement of best practices) and diffusion (the spread of the new technologies to other firms). Specifically, researchers found that not only did the best plants make rapid progress in emission reductions, but also the other plants caught up to such a high degree that they ultimately ended up lowering their emission intensities even more than the best plants.

- During the 1990s, as the demand for biofuels increased in Sweden, in part due to the energy and carbon taxing system, several new wood-handling technologies were introduced.

The finding that market-based instruments hasten innovation and diffusion is not, however, a universal finding. In some circumstances, pollution pricing may incentivize the exploitation of low-cost *existing* strategies (such as switching to lower carbon fuels) rather than stimulating the adoption of new technologies, thereby delaying their commercialization relative to other policies such as renewable portfolio standards. In these cases with carbon pricing, the stimulus to more fundamental innovation takes place over a longer time frame, once the existing cheaper opportunities are exhausted.

Conclusion

As a recent report from the National Academy of Sciences in the United States put it, "A carbon pricing strategy is a critical foundation of the policy portfolio for limiting future climate change" (NAS, 2010, p. 6). Carbon pricing is viewed as critical not only because it fosters the transition to a low carbon economy (and action—abatement—is ultimately cheaper than inaction—paying damages), but because it accomplishes that goal in a more cost-effective way due to the flexibility it provides businesses and citizens in making energy and related choices.

As this survey demonstrates, these policies have been around for a long time and this longevity provides a good basis for evaluating and refining the details of their implementation. This evidence demonstrates that not only can both forms of carbon pricing produce the desired reductions, but they can do so in a more cost-effective manner.

The old adage, "if it seems too good to be true, it probably is" certainly could apply to some descriptions of carbon pricing. The experience reviewed in this survey certainly does not reveal perfection. It does reveal, however, that imperfections can be addressed by better design. Carbon pricing is certainly not a utopian solution, but it seems to do the job and do it reasonably well.

References and Suggested Readings

Hedegaard, Connie, 2011, "Our Central Tool to Reduce Emissions," speech presented at the launch of Sandbag's report *Buckle Up! 2011 Environmental Outlook for the EU ETS*, European Parliament, Brussels, July 14.

National Academy of Sciences (NAS), 2010, *Limiting the Magnitude of Future Climate Change*, The Panel on Limiting the Future of Climate Change as part of the America's Climate Choices study (Washington: The National Academies Press).

Sterner, Thomas, and Bruno Turnheim, 2008, "Innovation and Diffusion of Environmental Technology: Industrial NO_x Abatement in Sweden under Refunded Emission Payments," *Ecological Economics,* Vol. 68, pp. 2996–3006.

For more information on some of the carbon pricing programs discussed in this chapter, see the following:

EU ETS: http://ec.europa.eu/clima/policies/ets/

U.S. Regional Greenhouse Gas Initiative: www.rggi.org.

Australia's carbon mitigation program: www.cleanenergyfuture.gov.au/clean-energy-future/our-plan.

For a discussion of the evidence on cost savings from emissions pricing programs over regulatory alternatives, see the following:

Tietenberg, Tom, 2006, *Emissions Trading: Principles and Practice*, 2nd ed. (Washington: Resources for the Future).

For evidence on the impact of emissions pricing on technological development, see the following:

Jaffe, Adam B., Richard G. Newell, and Robert N. Stavins, 2003, "Technological Change and the Environment," in *The Handbook of Environmental Economics*, ed. by Karl-Göran Mäler and Jeffrey Vincent (Amsterdam: North-Holland).

For more on general carbon taxes and environmental tax reforms, see the following:

European Environment Agency, 2005, *Market-Based Instruments for Environmental Policy in Europe*, 69, No. 8, http://reports.eea.europa.eu/technical_report_2005_8/en/EEA_technical_report_8_ 2005.pdf.

Sumner, Jenny, Lori Bird, and Hilary Dobos, 2011, "Carbon Taxes: A Review of Experience and Policy Design Considerations," *Climate Policy*, Vol. 11, pp. 922–943.

Glossary of Technical Terms and Abbreviations

Additionality. An emissions reduction is called "additional" if that reduction would not have occurred in the absence of a crediting (or other climate) program.

Afforestation. The conversion of former agricultural and abandoned croplands back into forests.

Air-capture technologies. These involve bringing air into contact with a sorbent material that binds chemically with CO_2 and extracts the CO_2 from the sorbent for underground, or other, disposal. Conceivably, if these technologies could be scaled-up globally (and funded publicly), they might be used to slow atmospheric GHG accumulations, although only in high-warming scenarios, given their high costs.

Biomass generation coupled with carbon capture and storage (BECS). A potential technology (as yet unproven) for capturing the CO_2 released when biomass is fired with other power generation fuels. This would result in negative emissions, given that biomass growth takes CO_2 out of the atmosphere.

BRIC countries. Refers to the group of large, rapidly industrializing countries—Brazil, Russia, India, and China—that are projected to account for a large share of the future global growth in GHG emissions.

British thermal unit (Btu). Measurement of energy based on heat content.

Carbon budget. Specifies a maximum allowable amount of CO_2 emissions that a country can emit, cumulated over a long period (say 10 years).

Carbon capture and storage (CCS). An (as yet unproven) technology for extracting CO_2 emissions from smokestacks and transporting them via pipelines to underground geological storage sites.

Carbon dioxide (CO_2). The predominant GHG. To convert tonnes of CO_2 into tonnes of carbon, divide by 3.67. To convert a price per tonne of CO_2 into a price per tonne of carbon, multiply by 3.67.

Carbon tax. A tax imposed on CO_2 releases emitted largely through the combustion of carbon-based fossil fuels. Administratively, the easiest way to implement the tax is through taxing the fossil fuels—coal, oil, and natural gas—on the basis of their carbon content.

CO_2 equivalent. The warming potential of a GHG over its atmospheric lifespan (or over a long time period) expressed in terms of the amount of CO_2 that would yield the same amount of warming.

Clean Development Mechanism (CDM). Under this program, emission-reduction projects in developing countries can earn certified emission reduction credits that can be purchased by industrialized countries to meet a part of their emission reduction targets under the Kyoto Protocol.

Clean energy standard. A proposed policy to lower the CO_2 intensity of the power generation sector through requiring a shift toward clean, or relatively clean, fuels (in effect, it provides similar incentives for fuel switching as would a CO_2 per kWh standard).

Common but differentiated responsibilities. A principle of the UN Framework Convention on Climate Change calling for developed economies to bear a disproportionately larger burden of mitigation costs (e.g., by funding emissions reduction projects in developing economies), given that they are relatively wealthy and contributed most to historical atmospheric GHG accumulations.

Computable general equilibrium model. A model of the whole economy that captures how changes in one market interact with other markets, the government sector, and (usually) choices over labor supply and investment.

Conference of the Parties (COP). The governing body of the UN Framework Convention on Climate Change that advances implementation of the convention through decisions taken at its annual meetings.

Credit trading. In cap-and-trade systems, credit trading allows firms with high pollution abatement costs to do less mitigation by purchasing allowances from relatively clean firms with low abatement costs. Similarly, in regulatory systems, credit trading allows firms with high compliance costs to fall short of an emissions (or other) standard by purchasing credits from other firms that exceed the standard.

Deforestation. Clearance of forests, mostly in tropical countries, and mostly for the preparation of land for use as pasture or crops.

Discount rate. In the present context, this mainly refers to the rate at which future climate change damages are discounted back to the present. There are two alternative notions of the appropriate discount rate. One is the *descriptive* rate, which is inferred from observations of people's actual behavior (e.g., saving versus consumption decisions over time, allocations of investment among more and less risky assets). The other is the *prescriptive* rate, which is based on a decision maker's judgment over how the well-being of future generations should be weighed against that of the present generation.

Downstream policy. This refers to an emissions policy imposed at the point where CO_2 emissions are released from stationary sources (primarily from smokestacks at coal plants and other facilities).

Emissions leakage. This refers to a possible increase in emissions in other regions in response to an emissions reduction in one country or region. Leakage could result from the relocation of economic activity, such as the migration of energy-intensive firms away from countries whose energy prices are increased by climate policy. Alternatively, it could result from price changes, such as increased demand for fossil fuels in other countries as world fuel prices fall in response to reduced fuel demand in countries taking mitigation actions.

Emissions standard. Sets an allowable emissions rate for producers of energy or energy-using products. For vehicles, this would be CO_2 per kilometer averaged across a manufacturers' sales fleet, while for a power generator, it would be CO_2 emissions per kilowatt-hour averaged across plants. Allowing firms to trade credits among themselves and across different periods of time is important for containing the cost of the policy.

Emissions trading system or scheme (ETS). A market-based policy to reduce emissions. Covered sources are required to hold allowances for each tonne of their emissions or (in an upstream program) embodied emissions content in fuels. The total quantity of allowances is fixed and market trading of allowances establishes a market price for emissions. Auctioning the allowances provides a valuable source of government revenue.

Energy Modeling Forum (EMF). Based at Stanford University, the EMF provides a forum for discussing modeling results related to energy and environmental policy.

Equilibrium climate sensitivity. A parameter summarizing the projected long-term climate response (i.e., warming) to a doubling of atmospheric CO_2 (equivalent) concentrations over their preindustrial levels.

Feebate. This policy imposes a fee on firms with emission rates (e.g., CO_2 per kilowatt-hour) above a "pivot point" level and provides a corresponding subsidy for firms with emission rates below the pivot point. Alternatively, the feebate might be applied to energy consumption rates (e.g., gasoline per kilometer) rather than emission rates. Feebates are the pricing analog of an emissions (or energy) standard, but they circumvent the need for credit trading (across firms and across time periods) to contain policy costs.

Feed-in tariff. This policy accelerates investment in renewable energy technologies by offering long-term, guaranteed-price contracts to renewable energy producers.

Fiscal cushioning. In the present context, this refers to adjustments to broader energy taxes or subsidies that offset some of the environmental effectiveness of a formal carbon tax. It would be potentially important to monitor (and perhaps penalize) fiscal cushioning in an international carbon tax agreement.

Forest management. In a climate mitigation context, this refers to management practices that affect the amount of carbon sequestered in a forest. These practices might include conversion of forestland to plantations, postponing timber harvests, planting trees rather than allowing natural regeneration, thinning trees and undergrowth to enhance forest growth, controlling forest fires and other disturbances, and fertilizing.

Geo-engineering. This refers to the (potential) use of technologies for altering the global climate system to counteract the effect of higher temperatures. Most prominently, these technologies include "solar radiation management," the deflection of incoming sunlight through shooting reflective particles into the stratosphere. Geo-engineering technologies may be very inexpensive to deploy, but could involve severe downside risks (e.g., radically altering precipitation patterns, overcooling the planet), and they do not address the problem of ocean acidification.

Gigatonne (Gt). 1 billion (10^9) tonnes.

Greenhouse gas (GHG). A gas in the atmosphere that is transparent to incoming solar radiation but traps and absorbs heat radiated from the earth. CO_2 is easily the most predominant GHG.

Green Climate Fund (GCF). This is a proposed fund to transfer money from developed countries to the developing world in order to assist the latter with climate adaptation and mitigation projects.

Integrated Assessment Model (IAM). A model that combines a simplified representation of the climate system with a model of the global economy to project the impacts of mitigation policy on future atmospheric GHG concentrations and temperature.

Interagency Working Group on the Social Cost of Carbon (SCC). A group of representatives from U.S. executive branch agencies and offices tasked with developing consistent estimates of the SCC for use in regulatory analysis.

Intergovernmental Panel on Climate Change (IPCC). The IPCC assesses the scientific, technical, and socioeconomic information relevant for understanding climate change. Its Fifth Assessment Report (AR5) is to be published in 2014.

International Civil Aviation Organization (ICAO). A specialized agency of the United Nations whose objectives include providing safe, secure, sustainable, and efficient global civil aviation while minimizing aviation's adverse environmental effects.

International Maritime Organization (IMO). Another specialized agency of the United Nations whose main purpose is to develop and maintain a regulatory framework for addressing environmental, safety, and other issues related to international shipping.

Kyoto gases. This refers to the six gases for which emission reduction pledges were made under the Kyoto Protocol. They include carbon dioxide (CO_2), methane (CH_4), nitrous oxide (N_2O), sulfur hexafluoride (SF_6), and two groups of gases, hydrofluorocarbons (HFCs), and perfluorocarbons (PFCs).

Kyoto Protocol. Under this protocol, 37 "Annex 1" or developed countries committed themselves to reducing CO_2 and five other GHGs to about 5 percent below 1990 levels by 2012 (China was not part of the protocol and the United States never ratified it).

Light detection and ranging (LIDAR). This refers to aerial photography used to measure forest volume.

Market failure. A situation where the private sector by itself would not make production and consumption decisions that would be efficient from society's perspective. The present focus is primarily on the market failure caused by excessive generation of GHG emissions that are not priced for their environmental damages. However, other market failures, such as those causing underinvestment in clean technologies (in spite of carbon pricing) are also focused on.

Negative emissions technology. A technology that results in a net reduction in atmospheric concentrations of GHGs (e.g., BECS, air-capture technologies).

Offset. A reduction in GHG emissions in other countries, or in unregulated sectors, that is made (and credited) in order to reduce the tax liability or permit requirements for emissions covered by a formal climate mitigation program.

Overshooting. In the present context, this refers to a transitory increase in atmospheric GHG concentrations over and above some long-term target before projected concentrations eventually fall back to the target.

Parts per million (ppm). Units for measuring the concentration of GHG molecules in the atmosphere by volume.

Radiative forcing. The difference between incoming and outgoing radiation energy (expressed in watts per square meter). An increase in radiative forcing tends to warm global temperatures. Radiative forcing changes with changes in incoming solar radiation, in atmospheric concentrations of GHGs (which prevent outgoing radiation), and in aerosols like suspended particulates (which deflect incoming radiation).

Reduced form model. A system of equations used to define the relationship between variables in a simplified form.

Reducing emissions from deforestation and forest degradation (REDD). This is an effort to create a financial value for the carbon stored in forests and thereby offer incentives for developing countries to reduce CO_2 emissions from deforestation and forest degradation. "REDD+" goes further and rewards forest conservation and management practices that sequester carbon.

Renewable portfolio standard. A regulation that requires that a minimum share of power generation comes from renewable sources such as wind and solar. These policies are difficult to justify on climate grounds alone, as other instruments (e.g., comprehensive carbon taxes) are far more effective at exploiting emission reduction opportunities across the economy.

Rental payment for CO_2. In the present context, this refers to an annual payment for carbon sequestered in forests.

Sigmoid growth function. A relationship describing the growth/age profile of trees. Growth initially increases with age up to a maximum; beyond that point, it then decreases with age.

Social cost of carbon (SCC). This refers to the net present value of damages (e.g., to agriculture, human health) due to the change in future global climate resulting from an additional tonne of CO_2 emissions in a given year. It is expressed in monetary units and usually reflects worldwide damages (rather than damages to a particular country).

Social welfare function. A representation of the overall well-being of a population, given their consumption and other factors like the quality of the environment.

United Nations Framework Convention on Climate Change (UNFCCC). This is an international environmental treaty produced at the 1992 Earth Summit. The treaty's objective is to stabilize atmospheric GHG concentrations at a level that would prevent "dangerous interference with the climate system." The treaty itself sets no mandatory emissions limits for individual countries and contains no enforcement mechanisms. Instead, it provides for updates (called "protocols") that would set mandatory emission limits.

Upstream policy. In the present context, this refers to an emissions pricing policy imposed approximately at the point where fossil fuels enter the economy (e.g., on refined petroleum products or at the mine mouth for coal).

Contributors

Valentina Bosetti
Fondazione Eni Enrico Mattei (FEEM) and Centro Euro-Mediterraneo per i Cambiamenti Climatici (CMCC), Italy

Carlo Carraro
University of Venice, Italy

Ruud de Mooij
Fiscal Affairs Department, International Monetary Fund

Robert Gillingham
Independent Consultant, formerly Fiscal Affairs Department, International Monetary Fund

Charles Griffiths
National Center for Environmental Economics, U.S. Environmental Protection Agency

Michael Keen
Fiscal Affairs Department, International Monetary Fund

Elizabeth Kopits
National Center for Environmental Economics, U.S. Environmental Protection Agency

Alan Krupnick
Resources for the Future, Washington, D.C., United States

Alex Marten
National Center for Environmental Economics, U.S. Environmental Protection Agency

Robert Mendelsohn
Yale University, Connecticut, United States

Chris Moore
National Center for Environmental Economics, U.S. Environmental Protection Agency

Steve Newbold
National Center for Environmental Economics, U.S. Environmental Protection Agency

Sergey Paltsev
Massachusetts Institute of Technology (MIT), United States

Ian Parry
Fiscal Affairs Department, International Monetary Fund

Rick van der Ploeg
University of Oxford, United Kingdom

John Reilly
Massachusetts Institute of Technology, United States

Roger Sedjo
Resources for the Future, Washington, D.C., United States

Brent Sohngen
Ohio State University, United States

Tom Tietenberg
Colby College, Maine, United States

Roberton Williams
University of Maryland and Resources for the Future, United States

Ann Wolverton
National Center for Environmental Economics, U.S. Environmental Protection Agency

Index

[Page numbers followed by *b, f, n* or *t* refer to boxed text, figures, footnotes or tables, respectively.]